GCSE 9–1
GEOGRAPHY
AQA REVISION & EXAM PRACTICE

Dan Cowling,
Philippa Conway-Hughes,
Natalie Dow and
Lindsay Frost

Authors
Dan Cowling, Philippa Conway-Hughes, Natalie Dow:
The challenge of natural hazards
Physical landscapes in the UK
Urban issues and challenges
Fieldwork and geographical enquiries

Lindsay Frost:
The living world
The changing economic world
The challenge of resource management
Issue evaluation

Editorial team Haremi Ltd
Series designers emc design ltd
Typesetting York Publishing Solutions Pvt. Ltd.
Illustrations York Publishing Solutions Pvt. Ltd.
App development Hannah Barnett, Phil Crothers, RAIOSOFT International Pvt Ltd and Haremi Ltd

Designed using Adobe InDesign
Published by Scholastic Education, an imprint of Scholastic Ltd, Book End, Range Road, Witney, Oxfordshire, OX29 0YD
Registered office: Westfield Road, Southam, Warwickshire CV47 0RA
www.scholastic.co.uk

Printed and bound in India by Replika Press Pvt. Ltd.
© 2017 Scholastic Ltd
2 3 4 5 6 7 8 9   7 8 9 0 1 2 3 4 5 6

British Library Cataloguing-in-Publication Data
A catalogue record for this book is available from the British Library.
ISBN 978-1407-17685-7

Note from the publisher:

Please use this product in conjunction with the official specification and sample assessment materials. Ask your teacher if you are unsure where to find them.

The marks and star ratings in the Exam Practice section of the book have been suggested by our subject experts, but they are to be used as a guide only.

Answer space has been provided, in the Exam Practice section of the book but you may need to use additional paper.

Colour versions of the images in the Exam Practice section of the book and the answers for the Check its! in the Revision Guide section and practice papers in the Exam Practice section are available online.

How to use this book

Inside this book you'll find everything you need to help you succeed in the new GCSE 9–1 AQA Geography specification. It combines revision and exam practice in one handy solution. Broken down into topics and subtopics, it presents the information in a manageable format. Work through the revision material first or dip into the exam practice as you complete a topic. This book gives you the flexibility to revise your way!

DO IT!

Activities to embed your knowledge and understanding and prepare you for the exams.

SNAP IT!

Use the Snap it! feature in the revision app to take a picture, film a video or record audio of key concepts to help them stick. Great for revision on the go!

EXAM-STYLE QUESTIONS

Exam-style questions for each subtopic ramped in difficulty.

NAIL IT!

Tips written by a subject expert to help you in the revision process.

Case study

Case studies that illustrate the ideas that you have learned.

Geographical skills

Tasks to develop your geographical, practical, numerical, statistical and mapwork skills.

STRETCH IT!

Questions or concepts that stretch you further and challenge you with the most difficult content.

CHECK IT!

Check your knowledge at the end of a subtopic with the Check it! questions.

REVIEW IT!

Consolidate your revision with the Review it! questions at the end of every topic.

★ STAR RATING ★

A quick visual guide to indicate the difficulty of the question, with 1 star representing the least demanding and 5 stars signposting the most challenging questions.

MARKS
(5 marks)

Each question has the number of marks available to help you target your response.

PRACTICE PAPERS

Full mock-exam papers to enable you to have a go at a complete paper before you sit the real thing!

Revision Guide contents

Topic 7

Topic 8

Exam Practice contents

THE CHALLENGE OF RESOURCE MANAGEMENT

HOW TO REVISE!

PLAN YOUR REVISION

Get ahead by planning your revision!

Work out the **time** you have available for revising.

Think about when you work at your best. Are you a morning or an evening person?

Allocate **MORE TIME** for the topics you struggle with.

Revision works best in **SMALL BURSTS**, so keep sessions **SHORT AND SWEET**!

Remember to allow time to **PRACTISE** applying what you have revised.

Use your **revision app** to put together a revision timetable.

LOOK AFTER YOURSELF

Help your brain by looking after your whole body!

Take regular **breaks** from revising – your brain needs time to digest information in order to retain it.

HOTEL

Keep **hydrated** by drinking plenty of water – dehydration stops your brain from working at its full capacity.

Regular **exercise** helps stimulate the brain and will help you relax.

Get plenty of **sleep**, especially the night before an exam.

EAT WELL and limit unhealthy snacks – your brain needs fuel for memory and concentration.

Find methods of **relaxation** that work for you throughout the revision period.

BE PREPARED!

Limit potential stress on the day of an exam by getting everything you need ready the night before.

30

Highlight key pieces of written information and use different colours to classify it.

DO IT!
Take an active approach to revision. Create **revision cards**, **lists**, **mind maps**, **flow charts** and **infographics** like this one.

Revise in pairs or small groups and deliver presentations on topics to each other.

Annotate and summarise your class notes and revision pages.

SNAP IT!
Use your phone to take pictures of your revision material so that you can **revise on the go**!

REMEMBERING INFORMATION

Use different methods to digest information!

Create a safe place to **store** all of your revision notes and keep everything together.

Use **mnemonics** or songs to help learn tricky bits of information.

RETRIEVING INFORMATION

Use the information you have remembered!

REVIEW IT!
Revisit the content you have already revised to confirm your understanding.

Practise applying your knowledge using exam-style questions.

READ THE QUESTION CAREFULLY!
Make sure you understand what you are being asked to do.

CHECK IT!
Make sure you know the material using simple **recall** questions.

Mark your work to help you see where you can **improve**.

FOR HIGH-MARK QUESTIONS, SPEND TIME **PLANNING** YOUR ANSWER!

Stick to the **TIME LIMITS** you will need to in the exam.

FINISH

The challenge of natural hazards

Natural hazards

THEEXAM!

- This section is tested in Paper 1 Section A.
- You must know natural hazards, tectonic hazards, weather hazards and climate change.
- You must know a named example of an earthquake or volcano, a tropical storm and an extreme weather event in the UK.

Natural hazards are naturally occurring physical phenomena, which pose a potential risk to human life and/or damage to property.

Hazard risk

Hazard risk is the chance or probability of being affected by a natural event. There are different factors that can affect risk:

- **Urbanisation**: densely populated areas are at greater risk from natural hazards.
- **Poverty**: in developing countries, people are forced to live in areas at greater risk, for example, on unstable slopes prone to floods and landslides.
- **Deforestation**: the removal of trees increases the likelihood of flooding and landslides.
- **Climate change**: as temperatures get warmer, there will be more tropical storms. Areas where it is wetter may become flooded and other areas could be drier with an increased risk of **drought**.

DOIT!

Categorise the following natural hazards into atmospheric, geological and hydrological hazards:

- volcanic eruptions
- earthquakes
- tropical storms
- tsunamis
- landslides
- floods.

CHECKIT!

1 Give a correct definition of the term 'natural hazard'.

2 Name the three categories of natural hazards.

3 a Why are some people more at risk from natural hazards than others?

  b Will people be more or less at risk of natural hazards in the future?

Tectonic hazards

Plate tectonic theory

1 The Earth's crust is split into a number of plates.

2 There are two types of crust: oceanic and continental crust.

3 Plates move due to heat deep within the Earth's core causing convection currents in the mantle, or due to slab pull and ridge push.

4 Tectonic activity at plate margins causes earthquakes and volcanoes.

 STRETCH IT!

Tectonic plate movement

Slab pull and ridge push is another theory that explains why tectonic plates move. At destructive plate margins, slab pull is where the denser oceanic plate sinks back into the mantle under the influence of gravity, and this pulls the rest of the plate with it. At constructive plate margins, ridge push is where the **magma** rises at plate margins. As the magma cools it becomes denser, which causes it to slide down, resulting in the plates moving away from each other.

 DO IT!

Study Figure 1, showing the distribution of earthquakes and volcanoes. Plan an answer to explain why earthquakes and volcanoes occur at the edge of plate margins.

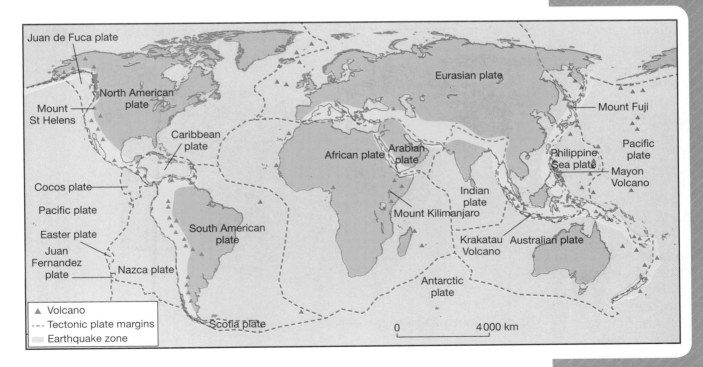

Figure 1 Distribution of earthquakes and volcanoes

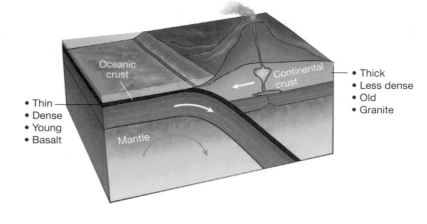

Figure 2 Oceanic and continental crust

Physical processes at plate margins

Constructive plate margins

At a constructive plate margin, two plates are moving away from each other. As the two plates move apart, magma breaks through the crust, causing earthquakes and volcanoes. The magma is very thin and runny, which allows the lava from an erupting volcano to travel long distances before cooling. This creates gently sloping shield volcanoes.

Sometimes the magma never reaches the surface, but can push up the crust to form ridges on the Earth's crust.

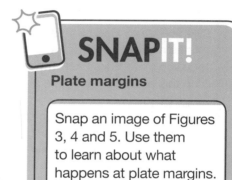

SNAPIT!

Plate margins

Snap an image of Figures 3, 4 and 5. Use them to learn about what happens at plate margins.

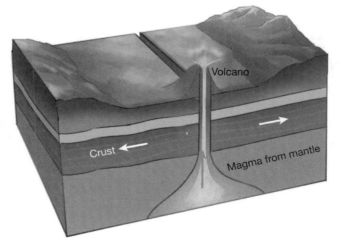

Figure 3 Constructive plate margin

Destructive plate margins

At a destructive plate margin two plates are moving towards each other. When an oceanic and a continental plate meet, the denser oceanic plate is subducted underneath the lighter continental plate. This creates deep ocean

DOIT!

Plate margins

Create a table summarising the differences between the three types of plate margin. Include:

- direction of plate movement
- tectonic activity (earthquake and/or volcanoes)
- type of plate (continental or oceanic)
- an example.

NAILIT!

Make sure you can describe and explain the relationship between plate margins and tectonic activity, such as why earthquakes and volcanoes happen at destructive plate margins.

trenches and fold mountains. The sudden release of the pressure that has built up from the plate being subducted causes strong earthquakes and, as the plate is being subducted, it allows magma to rise, resulting in very explosive volcanic eruptions. The magma is thick and forms steep-sided composite volcanoes.

When two continental plates meet, the crust is folded and uplifted to create fold mountains. This movement creates earthquakes but no volcanic eruptions.

Conservative plate margins

At a conservative plate margin two plates are moving past each other. Pressure builds up over many years and when the plate slips, strong, destructive earthquakes can be triggered.

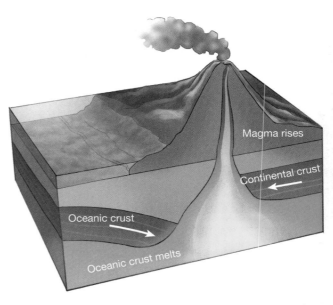

Figure 4 Destructive plate margin where an oceanic and a continental plate meet

Figure 5 Conservative plate margin

The effects and responses of tectonic hazards in countries of contrasting wealth

Primary effects

This is what happens directly as a result of an earthquake or volcano, such as injury, death and damage to roads and buildings.

Secondary effects

These are effects that occur indirectly as a result of an earthquake or volcano. They can happen hours, days or weeks later, such as landslides, fires and the spread of disease.

Case study

An earthquake in a developing country: Nepal earthquakes, 2015

An earthquake happened in Nepal on 25 April 2015, measuring a magnitude of 7.8. The epicentre was 80 km north-west of the capital, Kathmandu. A second earthquake in the same region occurred some days later on 12 May. The earthquakes happened because the Indo-Australian plate is colliding with the Eurasian plate – a destructive plate margin.

Primary effects

- Nine thousand people died and approximately 20 000 people were injured.
- Nearly three million people were left homeless.
- Following the earthquake, 1.4 million people needed food, even more needed water and shelter.
- Seven thousand schools were destroyed.
- Hospitals were overwhelmed and the international airport was congested, affecting the arrival of **aid**.
- The cost of damage totalled approximately US$5 billion.

Secondary effects

- An avalanche was triggered on Mount Everest, which killed 19 people.
- A landslide blocked the Kali Gandaki River, which resulted in people being evacuated in case of flooding.
- An avalanche in the Langtang region resulted in 250 people missing.

Immediate responses

- The UK, India and China provided rescue teams, water and medical support.
- The Disasters Emergency Committee (DEC) had raised US$126 million in **international aid** by September 2015.
- Half a million temporary shelters were set up.
- The United Nations (UN) and the World Health Organisation (WHO) distributed medical supplies to the areas that were most affected.

Long-term responses

- A post-disaster needs assessment was carried out and buildings were built to stricter codes.
- Areas where there were landslides were cleared and roads were repaired.
- The homeless had to be rehoused and 7000 schools were rebuilt.
- Mount Everest base camp was repaired so that the mountain could be reopened to climbers.

DOIT!

Nepal earthquakes, 2015

1 Sort the primary and secondary effects of the earthquakes in Nepal into the categories: social, economic and environmental. If you have used a different case study, use that as the basis for your answer instead.
2 Rank the effects, starting with the one you think is the worst.

 NAILIT!

Earthquakes in countries of contrasting levels of wealth

Earthquakes can have very different effects depending on what type of country they happen in. Remember why you often get worse impacts in a developing country.

Case study

An earthquake in a developed country: L'Aquila, Italy, 2009

An earthquake in central Italy happened on 6 April 2009, measuring a magnitude of 6.3. The epicentre was 7 km north-west of L'Aquila.

Primary effects

- More than 300 people died and 1500 people were injured.
- 67 500 people were left homeless.
- 10 000–15 000 buildings collapsed, including many churches and medieval buildings.
- The hospital was severely damaged.
- The cost of damage totalled approximately US$16 billion.

Secondary effects

- Landslides and rock falls caused damage to transport.
- The number of students at the University of L'Aquila has declined.
- Areas of the city were cordoned off due to unsafe buildings.

Immediate responses

- Hotels and tents were provided for those who were homeless.
- The Italian Red Cross searched for survivors – they also gave out water, tents, blankets and hot meals.
- The British Red Cross raised £171 000.
- Mortgages and bills for those affected were suspended.
- The EU granted US$552.9 million from its fund for major disasters to help L'Aquila straight away.
- The DEC did not provide aid, as Italy is a developed country.

Long-term responses

- A remembrance day procession is held on the anniversary of the earthquake.
- Residents paid no taxes during 2010.
- Students were exempt from university fees for three years.

DO IT!

L'Aquila earthquake, 2009

1 Sort the primary and secondary effects of the earthquake in Italy into the categories: social, economic, and environmental.

2 a Rank the primary effects, starting with the one you think is the worst effect.

  b Rank the secondary effects, starting with the one you think is the worst effect.

STRETCH IT!

The impact of earthquakes

Research shows that countries that are wealthier often suffer fewer impacts than developing countries. What evidence is there that the wealth of a country influences the extent to which it is affected by an earthquake?

DO IT!

Draw a revision poster showing the costs and benefits of living in an area that is prone to hazards.

Living with tectonic hazards

There are a number of reasons why people live in areas that are prone to hazards:

1. Large earthquakes and volcanic eruptions do not happen very often.

2. People do not have a choice if they are living in poverty.

3. Many plate margins run along coastlines, which are desirable places to live.

4. A lack of education in some areas means people do not know they are at risk.

5. Technological advancements have meant more effective monitoring of volcanoes and tsunamis.

6. Better building design reduces the risk for many settlements.

7. Volcanically active areas provide geothermal energy.

8. Extremely rich and fertile soils produced by volcanoes provide farming opportunities.

9. There are jobs in the mining industry for people living near volcanoes.

10. Many people visit as tourists, so tourism provides jobs and income.

Reducing the risk from tectonic hazards

There are four main strategies used to help manage the risk of tectonic hazards:

1. **Monitoring**: using scientific equipment to help detect the warning signs of a volcanic eruption or tsunami wave.

2. **Prediction**: scientists look at historical evidence and use monitoring equipment to help them make predictions about when and where a tectonic hazard may happen.

3. **Protection**: designing buildings that will withstand a tectonic hazard. This is more easily done for earthquakes than for volcanic eruptions.

4. **Planning**: authorities will identify areas most at risk from a hazard.

Monitoring

- **Volcanoes**: scientists will monitor volcanoes in a number of ways: they will use remote sensors to detect change in the volcano's heat and shape; seismometers to record any earthquake activity; tiltmeters to measure changes in the shape of the volcano as magma rises to the surface; and instruments to measure the gases being released.

- **Earthquakes**: these usually occur without any warning.

Prediction

- **Volcanoes**: prediction is based on monitoring the volcano. For example, before an eruption there is sometimes an increase in earthquake activity.

- **Earthquakes**: it is impossible to make accurate predictions, but scientists do use historical evidence to predict where an earthquake may be overdue.

Protection

- **Volcanoes**: it is harder to protect against violent volcanic explosions. Often, people will have to evacuate if a volcanic eruption occurs.

- **Earthquakes**: it is possible to create buildings that will withstand some earthquakes, and in many countries there are regular earthquake drills so that citizens know what to do.

Planning

- **Volcanoes**: there are hazard maps created for volcanoes to show the areas likely to be affected. This helps people to know which areas should be evacuated in the event of a volcanic eruption.

- **Earthquakes**: maps can be produced to show critical buildings, such as hospitals, in areas that are at high risk, so these areas can be protected.

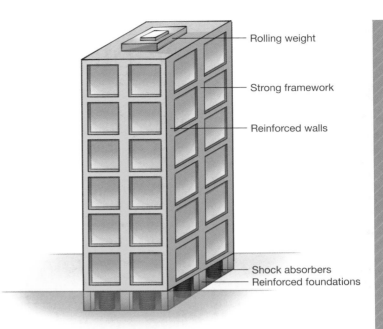

Rolling weight

Strong framework

Reinforced walls

Shock absorbers
Reinforced foundations

Figure 6 An earthquake-proof building

SNAP IT! Planning

Snap an image of Figure 6. Use this to help you remember how to make a building earthquake proof.

DO IT!

Managing tectonic hazards

1 Plan an answer to explain which method of risk management is the most effective.

2 How would hazard management be different between developing and developed countries?

CHECK IT!

1 Explain why earthquakes and volcanoes happen at plate margins.

2 Name the three different types of plate margin.

3 Explain why there are no volcanoes where two continental plates meet.

4 Give correct definitions for the terms 'primary effect' and 'secondary effect'.

5 Describe the effects of the L'Aquila earthquake or the earthquake in a developed country that you have studied.

6 'Earthquakes cause more damage in a developing country.' To what extent is this statement true?

7 State the four ways of reducing the risk from tectonic hazards.

8 Describe the benefits of living near a volcano.

9 Explain how buildings can be made earthquake proof.

Weather hazards

Global atmospheric circulation

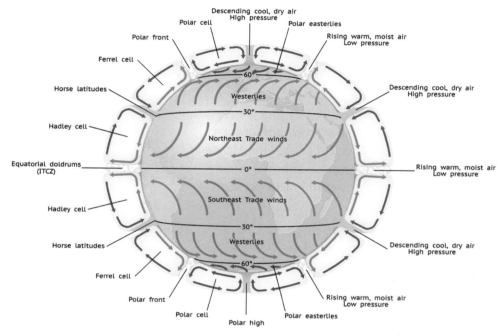

Figure 7 Global atmospheric circulation

Global atmospheric circulation helps us to understand the world climate zones and the pattern of global weather hazards.

1. Areas of high pressure are found where the air is sinking towards the ground, for example, at the North Pole. Winds on the ground move outwards from areas of high pressure.

2. Areas of low pressure are found where the air is rising, for example, at the equator. Winds on the ground move towards areas of low pressure.

3. Winds on the ground transfer heat and moisture from one area to another.

4. Patterns of pressure belts and winds are affected by the seasons and the tilt and rotation of Earth.

Tropical storms

Global distribution of tropical storms

Tropical storms develop in the tropics where there are areas of intense low pressure. They have different names depending on where they develop:

- Hurricanes in the USA and the Caribbean.

- **Cyclones** in South-East Asia and Australia.
- **Typhoons** in Japan and the Philippines.

Track and intensity of all tropical storms

TD Tropical depression
TS Tropical storm

| TD | TS | 1 | 2 | 3 | 4 | 5 |

Saffir-Simpson hurricane intensity scale

Figure 8 Global distribution of tropical storms

Formation of a tropical storm

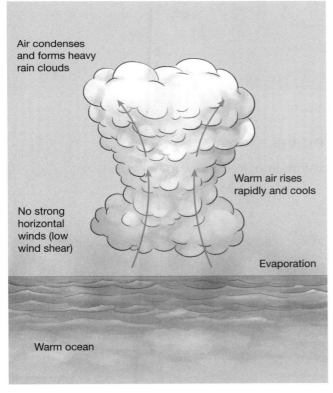

Air condenses and forms heavy rain clouds

Warm air rises rapidly and cools

No strong horizontal winds (low wind shear)

Evaporation

Warm ocean

Figure 9 Formation of a tropical storm

Write a paragraph describing the distribution of tropical storms. Use the following words in your notes:

- 5–15°
- north
- south
- equator
- tropics.

NAILIT!

Describing data in geography

When asked to describe a map or data in geography exams, always look for the overall pattern. Are there any anomalies? Include data to support your description.

1 Air is heated above the warm tropical oceans (27°C or above), causing it to rise rapidly.

2 Upwards movement of air draws up water vapour from the ocean's surface.

DOIT!

Draw a revision poster to show the formation of a tropical storm. Include an annotated diagram.

STRETCHIT!

The Coriolis effect

Research the Coriolis effect. How does the Coriolis effect influence the formation of tropical storms?

NAILIT!

Make sure you can explain how climate change could affect tropical storms in the future.

(3) Evaporated air rises and cools, causing it to condense to form large thunderstorm clouds.

(4) As air condenses it releases heat, which powers the storm and causes more and more water to be drawn up from the ocean.

(5) Several thunderstorms join together to form a giant spinning storm. It officially becomes a tropical storm when winds reach 63 km/h.

(6) The eye of the storm is created at the centre where air descends rapidly.

(7) As the storm moves across the ocean directed by prevailing winds, it develops in strength.

(8) When the storm hits land it loses its energy source (evaporated water), and friction with the land causes it to slow down and weaken.

Structure of a tropical storm

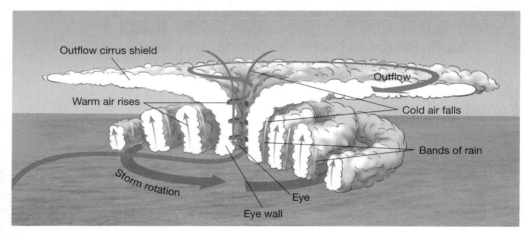

Figure 10 Structure of a tropical storm

Climate change and tropical storms

Climate change might affect distribution, frequency and intensity of tropical storms.

Distribution

The distribution of tropical storms could affect areas outside the current zone as sea temperatures increase.

Frequency

The frequency of tropical storms has increased, as six of the ten most active years since 1950 have been since the mid-1990s. The frequency in the future may decrease.

Intensity

Intensity of tropical storms could increase as sea surface temperatures increase.

STRETCHIT!

Tropical storms and climate change

Why are scientists not certain how climate change will affect tropical storms?

The effects of and responses to a tropical storm

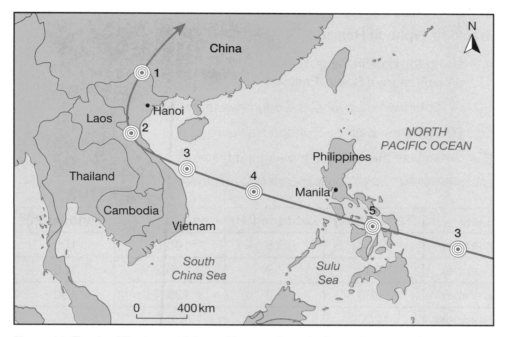

Figure 11 Track of Typhoon Haiyan. The numbers indicate the size of the storm along its path, with 5 being its peak

DO IT!

Primary and secondary effects

1 Sort the primary and secondary effects of Typhoon Haiyan, or the tropical storm you have studied, into the categories: social, economic and environmental.

2 Put the effects into rank order, starting with the one you think is the worst effect.

Case study

Typhoon Haiyan, Philippines, 2013:

Primary effects

- Approximately 6300 people died – mainly due to storm surge.
- Approximately 4 million people were displaced and 1 million homes were destroyed.
- The main airport terminal was destroyed.
- 30 000 fishing boats were destroyed.
- Widespread flooding.

Secondary effects

- Approximately 14 million people were affected.
- There were landslides and blocked roads due to flooding.
- Power supplies were cut off.
- A lack of shelter, water and food for many people led to an outbreak of disease.
- Jobs were lost; hospitals and schools were destroyed.
- Cost in damage totalled US$12 billion.

Immediate responses

- Food aid, water and temporary shelters were provided through international help.
- Aircraft and helicopters were provided to help with the rescue.
- Approximately 1200 evacuation centres, as well as field hospitals were set up.
- The Philippines Red Cross provided basic food aid.

Long-term responses

- The UN and other countries donated financial aid and medical support.
- Roads, buildings and bridges were rebuilt.
- Rice farming and the fishing industry were quickly restored.
- Homes were built in areas that are less likely to be affected by flooding.
- Cyclone shelters were built.

Practical skills

Tracking Typhoon Haiyan

1. Using the data in Table 1, plot out the path of Typhoon Haiyan on a blank map of South-East Asia using longitude and latitude.

2. On your own map label the countries that were affected.

3. Describe the path of Typhoon Haiyan.

4. When was the typhoon moving the fastest?

Date	Time (GMT)	Latitude (°N)	Longitude (°E)	Wind (mph)
5 November	1200	6.8	143.0	105
6 November	1200	7.9	136.2	160
7 November	1200	10.2	129.1	190
8 November	1200	11.8	120.6	155
9 November	1200	14.5	113.2	115
10 November	1200	19.3	108.1	85

Table 1 The path of Typhoon Haiyan, 2013

DOIT!

Reducing the effects of a tropical storm

1 Draw a spider diagram to explain the ways the effects of a tropical storm can be reduced.

2 How are the preparations for a tropical storm different in a developed country from a developing country?

Reducing the effects of a tropical storm

There are four main strategies used to help manage the risk of tropical storms.

Reducing the effects of a tropical storm	
Monitoring	• Satellites can monitor cloud cover and precipitation in areas where tropical storms are common.
	• Aircraft can fly into tropical storms at 3500 m to collect air pressure, rainfall and wind speed data.
Prediction	• All data is analysed by a supercomputer to predict the intensity and path of a tropical storm.
	• In the North Atlantic:
	○ Hurricane Watch advises of possible hurricane conditions.
	○ Hurricane warnings give advice on expected hurricane conditions and immediate actions that should be taken.

Reducing the effects of a tropical storm	
Protection	• Reinforce and strengthen windows, doors and roofs. • Construct storm drains in urban areas. • Construct sea walls to protect against storm surges. • Build houses on stilts to protect against flooding from storm surge. • Build cyclone shelters.
Planning	Planning mainly involves educating the community in what to do if a tropical storm hits the area: • Provide residents with disaster kits. • Always have fuel in car engine. • Knowledge of evacuation shelters. • Store loose objects.

Extreme weather hazards in the UK

Flooding

Heavy rainfall for an extended period of time can lead to river flooding. This is common in late winter and early spring, for example, the floods in southern England during the winter of 2013–2014.

Drought

Long periods of dry, warm weather can lead to drought in the UK. This affects rivers and reservoirs, which may dry up, affecting our water supplies.

Extreme cold

Periods of cold weather can bring frost, freezing and blizzard conditions. In the UK, this will affect transport, farming and industry.

Strong winds

The UK is sometimes affected by the tail end of a hurricane travelling over the Atlantic Ocean. This can disrupt power lines and cause trees to be blown down.

 STRETCHIT!

Research an extreme weather event in the UK. Find out about:

• its causes   • its effects   • the responses to it.

 DOIT!

1 Draw a revision poster to show extreme weather in the UK.

2 What secondary hazards could be caused by extreme weather?

UK extreme weather increase

Q Case study

A recent extreme weather event in the UK: Somerset floods, 2014

Causes

- Prolonged heavy rainfall in January and February 2014.
- There was 350 mm of rainfall in January and February (above average).
- Wet weather was brought by a series of depressions.
- High tides and storm surges followed.
- The River Tone had not been dredged for over 20 years.

Social impact

- Six hundred houses were flooded.
- Sixteen farms were evacuated.
- People were put up in temporary accommodation.
- Villages were cut off.
- Power supplies were cut off in some areas.

Economic impact

- Over £10 million worth of damage was caused.
- Approximately 14 000 ha of farmland were flooded for a month.
- Approximately 1000 livestock were evacuated.
- Part of the railway line was closed between Bristol and Taunton.

Environmental impact

- Sewage contaminated floodwater.
- Debris carried by the floods had to be cleared.
- Stagnant water had to be re-oxygenated before being pumped back into the river.

Immediate responses

- Villagers who were cut off had to use boats to go to school or the shops.
- A large number of volunteers helped the local communities.

Long-term responses

- Somerset County Council have put in place a £20 million flood action plan.
- Approximately 8 km of the Rivers Tone and Parrett have been dredged to increase capacity.
- Road levels have been raised in some areas.
- Flood defences have been built to protect vulnerable areas.
- River banks have been raised and strengthened.

DOIT!

Somerset floods

Draw a mind map to show the links between causes, effects and responses of the Somerset floods in 2014, or the UK extreme weather event you have studied if different.

STRETCHIT!

Some scientists link extreme weather events to climate change. Why can this be misleading?

There is evidence that some extreme weather events are becoming more frequent in the UK.

- **Precipitation**: there is evidence that precipitation could become more seasonal, but the overall total will be the same.

- **River flow**: the frequency and magnitude of winter river flooding has already increased. This could increase even further in the future.

- **Evaporation**: temperature in the UK has risen by 1°C and in the future temperatures will continue to increase, leading to more evaporation. This could result in more droughts in the UK.

Figure 12 Flooding on the Somerset Levels, 2014

CHECKIT!

1 Describe what low- and high-pressure systems are.

2 Explain why tropical storms do not occur at the equator.

3 Evaluate the different methods of reducing the effects of a tropical storm.

4 Explain why we get extreme weather events in the UK.

5 Discuss the different management strategies of reducing risk from flooding in the UK.

Climate change

Evidence for climate change

The **Quaternary period** is the period of time from 2.6 million years ago to the present day. During this time, temperatures have fluctuated. There have been cold periods, known as **glacial periods**, when parts of North America and Europe would have been covered in ice sheets, and warmer periods known as **inter-glacial** periods. Present-day temperatures are warmer than the rest of the Quaternary period. Since 1880, the average global temperature has risen by 0.85°C, noticeably since 1970.

Recent evidence for climate change

- Shrinking glaciers and melting ice.
- Rising sea levels:
  - When ocean water warms up, it expands in volume. This is known as thermal expansion.
  - Due to temperatures rising, more water flows into the oceans from glaciers and ice caps.
- Seasonal changes; for example, the UK growing season is lengthening.

Natural causes of climate change

1 Orbital changes

These are changes in the Earth's orbit:

- **Eccentricity**: the path of the Earth around the Sun. The Earth's orbit follows different elliptical paths. This happens on a 100 000 year cycle.
- **Precession**: the natural wobble of the Earth, which is on a 26 000 year cycle.
- **Tilt**: the Earth is currently at a tilt of 23.5 degrees. Over a period of 41 000 years, the Earth's tilt moves between 21.5 and 24.5 degrees.

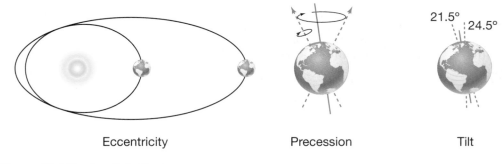

Eccentricity   Precession   Tilt

Figure 13 Milankovitch Cycles

DOIT!

Create a revision poster showing the evidence for climate change.

STRETCHIT!

Research how tree rings and ice cores can provide us with evidence for climate change.

NAILIT!

Make sure that you understand the three natural factors that affect climate change and whether they have a short- or long-term impact on the climate.

2 Solar output

- There are cyclical changes in solar energy linked to sunspots.

- A sunspot is a dark area that appears on the surface of the Sun.

- The number of sunspots increases from a minimum to a maximum and then back to a minimum over an 11-year period.

- The more sunspots there are, the more heat is given out (solar output). The fewer sunspots there are, the less heat is given out. This can reduce the Earth's temperatures.

3 Volcanic activity

- Volcanic ash can block out sunlight and so temperatures are reduced.

- When sulfur dioxide mixes with water vapour it converts into sulfuric acid. This reflects the Sun's radiation, reducing temperatures.

Human causes of climate change

SNAPIT!
Orbital changes

Snap an image of Figure 13 and use it to help you remember orbital changes. Think about how these changes affect the Earth's climate.

DOIT!

Orbital changes

Create a podcast to explain why orbital changes can influence the climate.

STRETCHIT!

Volcanic activity and climate

The Mount Pinatubo eruption in 1991 had a short-term impact on the climate. Research the effect of this eruption on the climate.

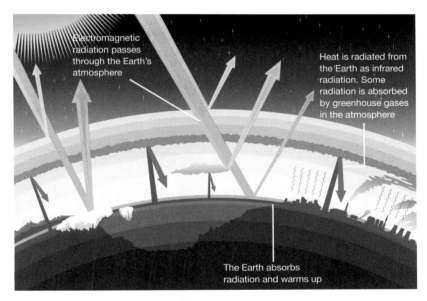

Electromagnetic radiation passes through the Earth's atmosphere

Heat is radiated from the Earth as infrared radiation. Some radiation is absorbed by greenhouse gases in the atmosphere

The Earth absorbs radiation and warms up

Figure 14 The greenhouse effect

SNAPIT! The greenhouse effect

Draw your own diagram of the greenhouse effect. Snap an image of it and use it to remind yourself of the effect. Can you explain how the greenhouse effect works and how it affects the climate?

DO IT!

Plan an answer to a question that asks you about how humans have contributed to climate change.

NAIL IT!

Make sure that you understand the difference between **human** and **natural** causes of climate change.

DO IT!

Look at the effects of climate change. Categorise these into 'effects on people' and 'effects on the environment'. Can you add any more to your list?

People adding more and more greenhouse gases to the atmosphere has led to the enhanced greenhouse effect. Most scientists believe this is the cause of recent global warming.

Fossil fuels

- The biggest proportion of greenhouse gases comes from fossil fuels.
- Burning fossil fuels releases carbon dioxide (CO_2) into the atmosphere.
- Fossil fuels are used in transport, industry and power stations to generate electricity.
- As population increases and people are becoming wealthier, more energy is used. This means more fossil fuels are burned.

Agriculture

- Produces large amounts of methane due to cattle and rice farming.
- As population increases, more food is required, especially in Asia where rice is a main part of the diet.

Deforestation

- Deforestation means there are fewer trees to take in carbon dioxide.
- Forests are carbon stores. When they are burned to clear an area, then the carbon dioxide that has been stored is released.

Effects of climate change on people and the environment

- An increase in flood risk due to increased levels of heavy rainfall.
- Rising sea levels will put pressure on sea defences and increase coastal flooding.
- An increase in extreme weather events, such as drought and heatwaves in the UK.
- An increase in crop yields in Europe, but a decrease in South-East Asia.
- Less heating needed in Northern Europe.
- Less ice in the Arctic, which could open up shipping routes.
- Wildlife affected, such as a decline in numbers of polar bears and seals in the Arctic.
- Warmer oceans could cause coral bleaching and reduce biodiversity in the Great Barrier Reef.

Managing climate change

Mitigation

Alternative energy production

Renewable energy sources, such as wind, solar and wave power, offer a way of reducing the amount of greenhouse gases being released.

Carbon capture

- Technology can replicate the way that the Earth stores carbon.
- Carbon dioxide is captured from emission sources and safely stored.

Planting trees

- Deforestation is a massive contributor to climate change.
- Reforestation increases carbon storage.

International agreements

- It is a challenge for all countries to be able to agree on targets to reduce climate change.
- Some countries can afford to mitigate against climate change more easily than others.
- Some countries are more responsible for causing climate change than others.

Adaptation

Change in agricultural systems

- Increasing irrigation where precipitation levels fall.
- Change in type of crop and time of year they are grown due to changes in temperature and precipitation levels.

Managing water supply

- Introducing more ways of conserving water and reducing waste, such as the use of water-efficient devices.
- Increasing water supply, such as desalinisation plants.

Reducing risk from rising sea levels

- Restoring mangroves to provide a natural sea defence in places such as the Philippines.
- Construction of sea defences, such as sea walls in the Maldives, to protect against rising sea levels.
- Building houses on stilts (examples in Miami, Holland and Suffolk).

NAILIT!

Make sure that you understand the difference between **mitigation** and **adaptation** strategies in managing climate change. Which one will be more effective in the long term?

DOIT!

Draw a revision diagram, summarising the different ways to manage climate change.

CHECKIT!

1 Describe the three causes of natural climate change.

2 Give a correct definition of the term 'greenhouse effect'.

3 Explain the difference between the greenhouse effect and the enhanced greenhouse effect.

4 Explain how the effects of climate change might vary in different parts of the world.

5 What is the difference between adaptation and mitigation?

6 Describe two ways to mitigate against climate change.

REVIEW IT!

The challenge of natural hazards

1 What is a natural hazard?

2 Give two ways in which you can reduce the effects of tropical storms.

3 Describe the global distribution of earthquakes and volcanoes.

4 Describe the benefits of living next to a plate margin.

5 Describe the formation of a tropical storm.

6 Assess the extent to which long-term responses are more significant than intermediate responses.

7 Explain why volcanoes are formed at destructive plate margins.

8 Explain how solar activity can affect global climate change.

9 Explain how humans contribute to an increase in carbon dioxide in the atmosphere.

10 Using a named example, explain the effects of an earthquake.

11 Using a named example, explain the effects of an extreme weather event in the UK.

12 'A developing country is likely to be more affected by an earthquake than a developed country'. To what extent do you agree with this statement?

The living world

Ecosystems

Scale of ecosystems

Ecosystems have a range of sizes:

- **Small-scale local ecosystems** such as ponds or sand dunes. Small-scale ecosystems are influenced by local factors such as rock type, soil characteristics and amount of water.

- **Large-scale global ecosystems** such as tropical rainforests, hot deserts or **tundra** (see Figure 5 on page 34). These large-scale global ecosystems are sometimes called **biomes**. Large-scale ecosystems are influenced by global factors such as **climate**: rainfall, temperature and seasons over thousands of years.

Case study

A small-scale UK ecosystem: Epping Forest, Essex

Epping Forest is an example of a small-scale UK ecosystem. It is a typical English lowland deciduous wood of great age. It was once used as a royal hunting forest and then pollarded. It has an area of just under 2500 ha – 70 per cent of its area is designated as a Site of Special Scientific Interest (SSSI) for its biological interest, and 66 per cent is designated as a Special Area of Conservation (SAC).

Components and interrelationships in the ecosystem

- The deciduous trees include oak, beech, hornbeam, birch and holly; deciduous trees lose their leaves for the colder months so that they are not damaged. Pollarding is the removal of the top branches of a tree to get new growth of denser branches and leaves, usually to increase the supply of wood.

- Pollarded trees have an unusual shape with numerous thick branches growing out of short trunks. This leads to many branches falling in stormy conditions creating lots of dead wood, which is perfect for **decomposers**. This helps decomposition of humus to add to the poor sand and gravel soils and supports birdlife such as woodpeckers.

- The oak trees support grey squirrels. Muntjac and fallow deer graze the forest floor along with rabbits. There are ten species of bat in the forest and also lizards, grass snakes and adders.

Balance in the ecosystem

- The Epping Forest Act (1878) stopped pollarding, allowing the trees to grow to their current shape blocking out much light from the ground and, therefore, reducing vegetation variety at ground and shrub level.

THE EXAM!

- This section is tested in Paper 1 Section B.

- You must know ecosystems and tropical rainforests.

- You must know **one** of hot deserts or cold environments (you do not need to revise both).

DO IT!

UK small-scale ecosystem

Create a revision card that has six bullet points to show the links between **producers**, **consumers**, decomposers, **food chain**, **food web** and nutrient cycling within a UK small-scale ecosystem.

Hint: do not be too general – include information and facts specific to the UK small-scale ecosystem you have studied.

- The deer are enclosed within the forest to reduce collisions with vehicles on the busy roads that surround the forest. The deer eat a lot of the ground-level plants and damage trees.

- The forest is surrounded by urban areas, including part of Greater London, so there are pressures from human activities such as walking, horse riding and mountain biking.

- An SSSI assessment report found the forest air to be of poor quality causing damage to the older trees. Deer management is also needed as they are eating too much vegetation.

- Droughts (such as those during 1976–77) and storms (such as the Great Storm of 1987) have an impact on the balance within the forest ecosystem, causing the death of trees.

Interactions within ecosystems

The parts of an ecosystem can be divided into abiotic and biotic:

- Abiotic means the non-living parts of an ecosystem, such as soil and water.
- Biotic means the living parts of an ecosystem, such as plants and animals.

The living and non-living parts of an ecosystem interact (or interrelate) through several systems, such as the nutrient cycle and water cycle. The living parts of an ecosystem interact within a food web (consisting of many interlinked food chains).

Links within an ecosystem are often in both directions, for example, leaves from a tree fall to the ground, decompose and add nutrients to the soil (forming humus), and then the tree absorbs these nutrients (dissolved in water) through its roots.

Food web and food chain

Food is the way in which energy is passed from one level of an ecosystem to the next; these levels are called trophic levels (see Figures 1 and 2).

- A food chain is where living things are placed in a line of what eats what (see Figure 1).

- A food web is where many food chains link together within an ecosystem, with certain living things being the food for several creatures (see Figure 2) or one creature eating lots of things.

SNAPIT! Food chains in a food web

Take an image of Figure 2 and compete with a friend to see who can spot the most food chains. Then think about what would happen to other parts of the ecosystem if the caiman became extinct.

NAILIT!

Ecosystem interactions

Make sure that you can describe and explain the interactions within an ecosystem using the correct terms, such as: food chain, food web, biotic, abiotic, producer and (primary and secondary) consumers.

DOIT!

Food webs

With a friend, each create four multiple-choice questions that test understanding of the terms associated with food webs. Test your friend. Ask them to test you with their questions.

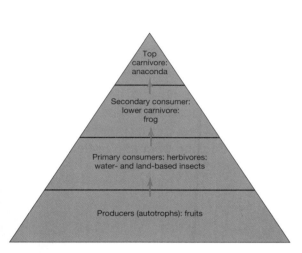

Figure 1 A tropical rainforest food chain

Figure 2 Part of a tropical rainforest food web

Producers, consumers and decomposers

1 **Primary producers** are sometimes known as **autotrophs** – these are plants that automatically convert sunlight, water and minerals through the process of **photosynthesis** into energy and starches so that they can grow.

2 Plants are the basis for all the other levels of creatures in an ecosystem, as they are the food (leaves, nectar, seeds, fruits, nuts) for consumers.

3 Consumers are either herbivores/primary consumers (creatures that eat plants), carnivores (creatures that eat other creatures), or omnivores (creatures that eat both plants and animals).

4 When producers or consumers die, they accumulate on or in the ground where decomposers (such as bacteria or fungi) break them down into basic minerals, which become nutrients within an ecosystem.

Nutrient cycle

- Mineral **nutrients** accumulate in soil, dissolve in rainwater and are absorbed by plants through their roots. Some important minerals are carbon, nitrogen and phosphorus.

- Nutrients help plants to grow and the minerals are then passed on to consumers through the food web. The cycle is completed when leaves fall to the ground to become litter, which decomposes, or waste products from animals are added to the soil along with their bodies when they die. In this way, nutrients are moved around and around (see Figure 3) unless there is interference from human activity.

- Therefore, nutrients are stored in living things (**biomass**), soil and litter (dead leaves and twigs on the ground).

> ## NAILIT!
>
> ### Humus
>
> Decomposed vegetation in the soil is known as humus. This is an important source of nutrients for plants and helps to bind soil together by holding moisture.

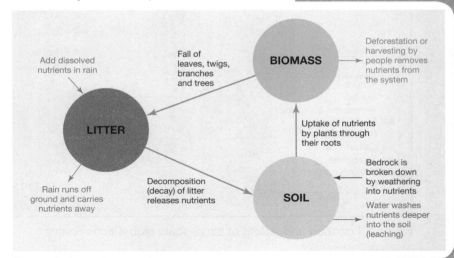

Figure 3 A nutrient cycle

Balance and change

A natural ecosystem develops a balance between all of its parts (biotic and abiotic) over a long period of time. In this way, ecosystems remain stable (change very little) unless there is a significant event to cause change.

- Change caused by human activity: this includes the removal of natural vegetation and the alteration of soils to create farmland (see Figure 4), as well as climate change caused by emissions of greenhouse gases over the last 200 years, which is making south-east England drier and hotter. The historical extinction of carnivores in the UK means that deer have no predators and can multiply and consume more vegetation.

- Change caused by physical conditions: small-scale UK ecosystems may face extreme weather conditions such as drought (e.g. in 1976–77) or a great storm (e.g. in 1987), and diseases such as ash dieback impact on the rest of the ecosystem by changing the food supply in the food web.

Figure 4 Deforestation in Brazil: an aerial view of a large soy field eating into the tropical rainforest

DO IT!

Tropical rainforest nutrient cycle

Draw an annotated diagram to show how nutrients are transferred (cycled) in a tropical rainforest ecosystem.

DO IT!

Change in a tropical rainforest

Make your own sketch of Figure 4. Label all the differences between the forest and deforested areas that you can identify.

Distribution and characteristics of large-scale global ecosystems

There is a pattern of large-scale global ecosystems (biomes) largely based on distance from the equator (latitude) (see Figure 5). This pattern is modified by mountain ranges and by distance from the sea. Tropical to Mediterranean

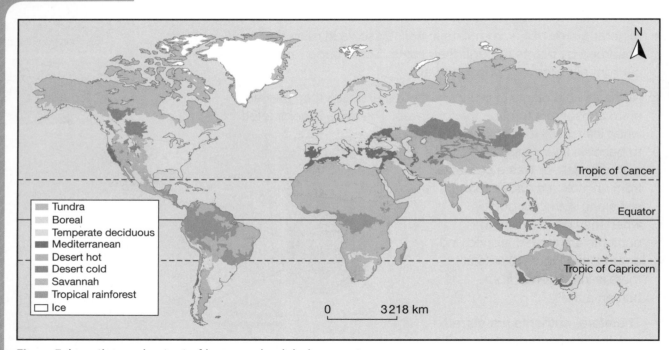

Tundra
Boreal
Temperate deciduous
Mediterranean
Desert hot
Desert cold
Savannah
Tropical rainforest
Ice

Tropic of Cancer

Equator

Tropic of Capricorn

0   3218 km

Figure 5 Location and extent of large-scale global ecosystems ecosystem patterns are clearest in Africa, as there is nearly a mirror image north and south of the equator in this continent (see Figure 8 on page 37). The mid-latitude to polar ecosystem pattern is clearest in the northern hemisphere because there is more land.

Tropical rainforests

1. Tropical rainforests are located mostly between the tropic lines either side of the equator: the Tropic of Cancer in the northern hemisphere and the Tropic of Capricorn in the southern hemisphere. The largest concentrations are in South America (e.g. Brazil), West and Central Africa (e.g. Democratic Republic of Congo) and South-East Asia (e.g. Indonesia).

2. These areas are very hot and wet for most of the year, although some exist where there is a dry season. Average temperatures are between 25° and 30°C and precipitation is between 2000 and 3000 mm a year. It is very hot because the Sun's energy is concentrated on a small surface area and in a small volume of atmosphere (see Figure 7 on page 36). It is wet because as the air is heated it rises and cools so that the evaporated water condenses into large clouds (convectional rain). These clouds create heavy tropical downpours, especially in the afternoons and in the cool of evenings. The rising air is the start of the Hadley convection cell (Figure 6 on page 36) at the intertropical convergence zone (ITCZ).

3. The vegetation is based on layers of trees, which thrive due to the ideal combination of sunlight, heat and water.

Hot deserts

1. Hot deserts are found either side of the Tropic of Cancer and Tropic of Capricorn where it is hot and dry (over 30°C, with less than 250 mm rainfall a year). At night, temperatures drop dramatically as there are no clouds to trap the heat of the day. This creates a large diurnal temperature range.

2. Hot deserts are found where air sinks on a large scale as part of the global atmospheric circulation, such as the Sahara and Kalahari deserts in Africa. This occurs where the air in the Hadley convection cell moves from high altitude to ground level. This air is dry because most of its moisture was left behind near the equator (Figure 6). As the air sinks, it warms and any moisture left is in the form of water vapour, so there are few clouds. It is also stable and without turbulence: any humid air that exists is not lifted up to cool.

SNAPIT!

Location of large-scale global ecosystems

Snap an image of Figure 5 and use it to remind yourself of where the major biomes of the world are located. When you are doing this, also think about **why** they are in these locations.

STRETCHIT!

Influences on biome location

Mountains are cooler and wetter than other areas and so ecosystems change with altitude. Warm ocean currents, such as the Gulf Stream, keep land areas warmer and wetter, while cold currents keep areas drier as there is less evaporation.

NAILIT!

Location and distribution

Make sure that you know the difference between location and **distribution**. Location is the precise position on the Earth's surface of a geographical feature, such as a tropical rainforest, often using latitude and longitude, or place names. Distribution is the amount and spatial arrangement (or spread) of a geographical feature in a region or different parts of the world.

③ There are not many plants because of the lack of moisture and poor soil – only a few that have adapted, often by storing water – and animals usually only come out at night.

Tundra

① Tundra ecosystems are found in zones in and around the **polar regions**. There is only a small amount of tundra in the southern hemisphere because there is limited land at the correct latitude. However, there are large areas in the northern hemisphere – especially Alaska (USA), northern Canada and northern Russia along the shores of the Arctic Ocean.

DOIT!

Hadley convection cell

Create a mnemonic to help you remember what happens in the Hadley convection cell and how it produces the location of tropical rainforests and hot deserts.

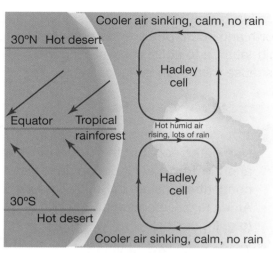

Figure 6 Links between Hadley convection cell and tropical rainforest and hot desert ecosystem locations

NAILIT!

Atmospheric circulation

Make sure that you understand that the lower layer of the Earth's atmosphere has very large convection cells. When air rises, unstable conditions are created which lead to rain. When air sinks, stable conditions are created and it is dry.

② Temperatures are below 0°C for most of the year and peak at about 10°C in the short summer. Precipitation is below 250 mm because evaporation rates are low and the air is stable because it is sinking. It is so cold here because the Sun's energy is spread out over a larger surface area and larger volume of atmosphere due to the Earth being the shape of a sphere or ball (see Figure 7). Also, north of 66.5° the Sun never rises during the middle of winter.

③ There aren't many plants because it is so cold. Plants grow low to the ground to avoid bitterly cold winds and animals have adapted to the cold conditions, often with thick fur.

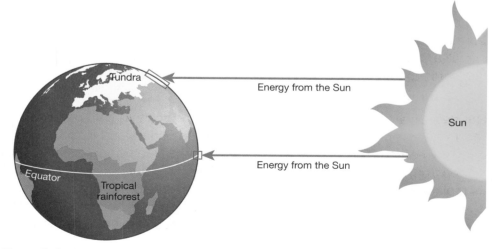

Figure 7 Reasons for difference in heat energy from the Sun

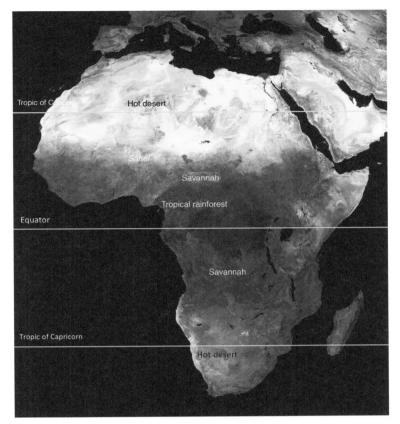

Figure 8 Composite satellite image of the continent of Africa showing ecosystem locations through colour (equator and tropic lines included)

NAILIT!

Remember that it is **not** hotter at the equator because it is closer to the Sun. The distance difference between pole to Sun and equator to Sun is very tiny. The difference in temperature is caused because the Earth is shaped like a ball and the Sun's energy is concentrated at the equator and more spread out at the poles (see Figure 7).

DOIT!

African biome locations

Look at Figure 8. Close your eyes and remember the image, concentrating on the colours and the equator and tropic lines. Describe the pattern of biomes to a friend, or close your book and try drawing the pattern of biomes from memory.

CHECKIT!

1 Give an example of a biotic factor.

2 Give the definition of the term 'abiotic'.

3 Give two differences between a food chain and a food web.

4 Describe how nutrients are cycled through an ecosystem.

5 Give the location of the tundra global ecosystem.

6 Describe the climate of the hot desert global ecosystem.

7 Explain the location of the tropical rainforest global ecosystem.

8 Describe two ways in which people may change the balance within a large-scale global ecosystem.

9 Explain the role of the Hadley convection cell in influencing the location of hot deserts and tropical rainforests.

Tropical rainforests

Characteristics of tropical rainforests
Physical characteristics

1. Tropical rainforests have a clear structure with different layers of vegetation. There are the tallest isolated trees like the kapok that grow over 50 m tall; these are known as **emergent trees**. Below them, there are at least two levels of trees which form **canopies** from their dense green leaves. The trees in the main canopy are about 35 m tall. Below them, there is a **shrub layer** (with young trees) where enough sunlight gets through the **canopy** layers (see Figure 9). There is very little vegetation at the ground level as it is quite dark, which means that there is less food for animals at ground level. Most food all year round is found in the canopy layers and, therefore, creatures, such as spider monkeys, have adapted to living high in the trees. Ground-level vegetation is only dense where there is sunlight, such as along riverbanks.

Figure 9 The structure of the Australian tropical rainforest

2. The **growing season** for plants is 12 months, as it is always hot and wet – ideal growing conditions. With such a wealth of food it is not surprising to find that the tropical rainforest is the most biodiverse ecosystem on the planet. It is home to about 50 per cent of all species, with many found only in a small area.

3. The hot, wet conditions also speed up nutrient cycling. The largest store of nutrients is in the large trees (biomass). The litter from these decomposes quickly in the top soil layer. Then the shallow roots of the plants absorb them quickly before they are washed away.

4. Soils lack nutrients because the nutrients are recycled quickly back into the plants or they are washed deep into the soil by the heavy tropical rain.

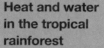

NAIL IT!

Heat and water in the tropical rainforest

Make sure that you understand the influence of heat and water on the growth of plants, especially trees, in the tropical rainforest and their influence on nutrient cycling.

DO IT!

Tropical rainforest ecosystem

Create a podcast to explain why the tropical rainforest has the greatest **biodiversity** of all the large-scale global ecosystems on Earth.

Interdependence

- As with all ecosystems, each part of a tropical rainforest is linked with other parts and they depend on each other. For example, trees provide a place for creatures to live and food for them; in return, these creatures spread the seeds of the trees to new growing places.

- The nutrient (or mineral) flow in a tropical rainforest is fast because it is so hot and wet and, therefore, chemical reactions are faster. Leaves and branches that fall on to the ground decompose quickly and the nutrients enter the top layer (horizon) of the soil; the roots of the trees are shallow and spread out so that they can quickly absorb the minerals before they are washed deep into the soil by the heavy tropical rain (a process called leaching).

- The soils (laterites) are not very fertile. They lack nutrients because of leaching. Most of the nutrients in the tropical rainforest ecosystem are stored in the vegetation (e.g. trees 50–60 m tall). If the trees are removed, then the nutrient cycle is interrupted and problems occur as fewer nutrients are left within the ecosystem.

- Trees transpire moisture through their leaves, which then evaporates into the air. This is called evapotranspiration and creates high humidity. When the air temperature cools in the evenings, night condensation takes place forming large clouds, which then cause rainfall. Without this transpiration, the tropical rainforest areas would be drier.

- People have lived in the forest for thousands of years, using the resources that the forest provides. These native (indigenous) peoples developed a balance with the natural systems. However, modern people have the motive and ability to damage the tropical rainforest on a much larger scale.

Adaptation of plants and animals

Plants have adapted by:

- growing tall to reach the sunlight

- having large leaves that pivot on their stalks to capture and follow sunlight

- having waxy leaves with drip tips, so that the heavy rainfall runs off the tree quickly and mosses and other small plants can't grow on them

- having buttress roots that give stability at the base of the tallest trees

- having shallow root systems to obtain nutrients from the top layer of soil before they are washed away (process of leaching).

Animal life has adapted in the following ways:

- Monkeys and similar mammals are able to climb and swing between trees to find food.

- Gliding snakes flatten themselves into a ribbon to be able to glide between trees and have backward pointing scales to be able to climb trees.

SNAPIT!

Tropical rainforest structure

Snap an image of Figure 9. Use it to remind yourself of the four different levels within the rainforest and what each layer is like. Can you explain the natural processes happening in the main canopy layer and at ground level?

NAILIT!

Make sure that you understand that adaptations are designed to increase the chances of survival of plants and animals. For example, why do the trees grow so tall or why do some squirrels glide from one tree to another?

NAILIT!

Make sure that you understand the term 'biodiversity'. This means the number and variety of living organisms (species) within a given area. It may also be used to refer to the variations between biomes, ecosystems and species.

DOIT!

Plan an answer to a question that asks you about the problems created for an ecosystem and people by a loss of biodiversity.

Remember that the question is asking about problems created in an ecosystem **and** problems that affect people.

- Flying squirrels have skin between their legs and arms and bodies so that they can jump and glide between trees.

- Insects and small animals have developed camouflage to be harder to spot by predators.

- Frogs lay their tadpoles in pools of water trapped in flowers in the canopy layer.

Biodiversity issues

The tropical rainforest is the most biodiverse ecosystem on Earth, partly because it has been relatively unaffected by past changes to the global climatic cycles. Living things have therefore had lots of time to evolve. Each small area is home to different living things. Therefore, even the destruction of a small area can cause the extinction of plants and animals and biodiversity is lost. It is likely that many living things have become extinct without scientists even finding them.

Loss of biodiversity may cause issues such as:

- Extinction of species endangers others in the food web.

- Loss of genetic material from which to find medicines or new food resources.

- An unbalanced ecosystem makes the rainforest less resistant to change, especially sudden changes.

Impacts of deforestation

Cutting down the tropical rainforest may have serious consequences at all scales – local to global (see Figure 10):

- Locally, the soil may be exposed by **deforestation**, enabling the rain to wash it away (**erosion**) and leaving natural plants or farmers' crops nothing to grow in.

- Regionally, the climate may become drier because with fewer trees, the rate of evapotranspiration is reduced. Therefore, the air is less humid and rainfall decreases, making it even more difficult for plants to grow.

- Nationally, a country may gain money from selling resources obtained from tropical rainforest areas, but eventually these may be used up leaving fewer ways of making money.

- Globally, the rate of absorption of carbon dioxide is reduced as there are fewer trees to use the gas during photosynthesis, which means that more carbon dioxide stays in the atmosphere, contributing to global warming.

Rates of deforestation

1 Humans have always cut down the tropical rainforest. Indigenous tribes made their homes from timber and cleared small patches of land to grow crops, for example. However, this had little long-term impact and was **sustainable**.

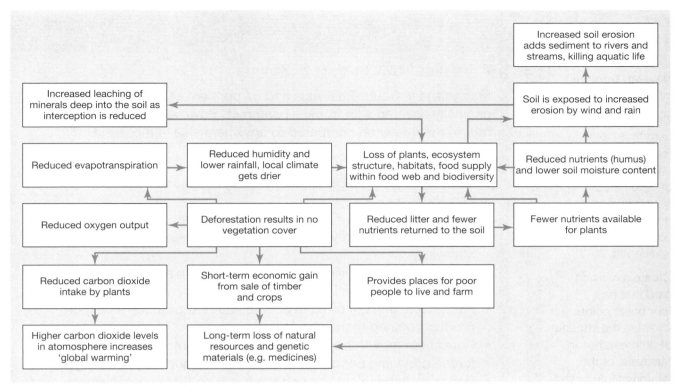

Figure 10 Impact of tropical rainforest deforestation

2 Modern humans with technology, such as large machinery, have cut down much larger amounts of forest. There are pressures from growing populations who need somewhere to live and farm, and there are pressures for making money so that a country can develop. Large companies can make money and people can move out of poverty.

3 Deforestation rates have varied over time (see Figure 11) depending on economic influences, such as demand for palm oil, soya, beef, open-cast mining of metals and supplies of **hydro-electric power (HEP)**. However, if current rates continue, the tropical rainforest could be all gone within 100 years.

Figure 11 Deforestation in the Amazon tropical rainforest, Brazil

DO IT!

Study the following data giving the original size of the Amazon tropical rainforest and its size now.

Original area of Amazon tropical rainforest: $4\,100\,000\,km^2$

Area of Amazon tropical rainforest in 2015: $3\,300\,000\,km^2$

Calculate:

1 The area of the forest that has been lost (in km^2).

2 The percentage of the forest now remaining.

DOIT!

Malaysian tropical rainforest

1 Create a revision card that has four bullet points showing the causes of deforestation in Malaysia, or the tropical rainforest you have studied if it is different.

2 Create a revision card that has four bullet points showing the impacts of deforestation in Malaysia, or the rainforest you have studied.

NAILIT!

Extent of tropical rainforest

Remember that the tropical rainforest is in South America, West and Central Africa and South-East Asia. Within these world regions the tropical rainforest overlaps several countries. For example, the Amazon rainforest is mostly in Brazil, but it is also found in Bolivia, Peru and Colombia. In South-East Asia, tropical rainforests are found in Indonesia, Malaysia and northern Australia.

Case study

A tropical rainforest: Malaysia

Malaysia is in South-East Asia and 67 per cent of the country was originally covered with tropical rainforest. However, it has the fastest rate of deforestation compared to anywhere else in the world.

Causes of deforestation

- Logging to get tropical woods for export, which brings money into Malaysia.
- Hydro-electric power (HEP) stations have been built with dams and reservoirs. These provide electricity for businesses and industries in Malaysia.
- Mining of tin and drilling for oil and gas take place in the rainforest so that Malaysia can sell products and improve energy security.
- Plantations of palm oil have been created to produce an export product to bring money into Malaysia.
- Population growth has encouraged settlements and farming in the forest, providing people with a place to live and make a living.
- Roads have been cut through the forest so that companies can get to the logging areas, mines, energy projects and new settlements.
- Forest fires caused by 'slash and burn' methods of clearing the forest for farming have spread and caused considerable damage.

Impacts (negative and positive) of deforestation

- Biodiversity is reduced and wildlife such as orang-utans are threatened by loss of habitat.
- Soils are exposed to erosion once the forest structure is destroyed.
- The extent of climate change is increased as the forest's ability to absorb carbon dioxide is reduced.
- Transpiration is reduced, so local areas become drier.
- Jobs are created by the primary and secondary industries based on forest products and tertiary jobs are linked to exporting the products.
- Malaysia earns money from timber, metal and energy exports, which can help develop the country.
- Water and air pollution result from the forest activities.
- Soils may be degraded so much that the land cannot be used for anything.
- An unsightly rainforest reduces ecotourist numbers.

Sustainable management

Ways of decreasing deforestation rely on organisations, people and governments working together – this is not easy to achieve.

Value of tropical rainforests

People

- For indigenous peoples, the tropical rainforests are their home, providing resources for shelter, food and medicines.

- For people with economic motives, there is a timber resource, agroforestry crops, other foods, medicines, scientific, aesthetic and ecotourism value.
- The tropical rainforest also helps to balance the amounts of oxygen and carbon dioxide in the atmosphere, providing the oxygen that people need to breathe and storing carbon.

Natural environment

- It is known that the tropical rainforest ecosystem has an influence on local climate and scientists think that it may have an important influence on global climate.
- It is the oldest biome and has the greatest biodiversity by far, with complex food webs and unique examples of evolution on Earth.
- It acts as a store of carbon within the carbon cycle, balancing the gases in the atmosphere.

Sustainability strategies

Selective logging and replanting

- Selective logging involves choosing carefully where to extract timber and the type of timber to extract. For example, mahogany has been a popular wood in developed countries because of its colour, but these trees grow on their own and to get to them a lot of other trees are cut down. So an approach using other more numerous types of tree instead would reduce needless deforestation.
- As well as selective logging, the deforested area could be immediately replanted with the same trees as those removed, which would maintain the forest area and biodiversity.

Figure 12 Tree replanting

NAILIT!

Carbon cycle

Carbon is found in several forms. The Earth's natural systems move carbon between the biosphere, atmosphere, oceans and rocks, as well as storing it in these places for a long or short time.

DOIT!

Study all the information on sustainable strategies for using the tropical rainforest. Decide which you think is the best approach. Write down three reasons why you think this.

SNAP**IT!** Replanting

Find an image of tree replanting in a tropical rainforest and Snap it! Use this to remind you of the steps necessary to persuade people to carry out replanting of tropical rainforests.

Conservation and education

- Conservation means looking after the forest so that it is not damaged. It does not necessarily mean preventing human use of the forest, but it does involve protection, such as national parks or wildlife reserves where human activities are restricted by law. It can also include 'buffer zones' where carefully monitored human activity takes place to ensure that any deforestation is under control.

- People in all countries can be educated about the value of the tropical rainforest and also about farming and logging methods that cause minimal damage without affecting their income or access to resources.

Ecotourism and international agreements

- **Ecotourism** is popular because people wish to see undisturbed natural areas. Local people living in tropical rainforests can gain jobs and income by being guides, offering accommodation or selling local craft products. The local people can therefore make more money from visitors than they could by destroying the rainforest.

- All countries are linked, especially economically. Some developed countries may influence the developing countries that have tropical rainforest by helping with monitoring deforestation, or through directly funding international organisations or programmes, for example, reducing emissions from deforestation and degradation (REDD).

DO**IT!**

Conservation of a tropical rainforest

Using Figure 13, draw a simple flow diagram to show the steps needed to reduce deforestation using the REDD principles.

Recover degraded land where pressures from people have been high, replant variety of trees to restore biodiversity and reduce soil erosion.

Stop forest destruction Modern machinery used to clear large areas of forest. Burning of forest adds to damage. Biodiversity lost.

Use forest sustainably Indigenous people living in harmony with nature. Forest structure maintained.

Maintain forest peatland Prevent encroachment into wetland habitats. Make nature reserve.

Figure 13 A REDD sponsored project

Debt reduction

Many developing countries, such as Brazil, borrowed money in the past to help build expensive schemes to aid development of the country. However, these countries were often unable to pay back the money and are now in debt. One type of international agreement is a 'debt-for-nature swap' or **debt reduction**, where some of the money owed is written off (deleted) in return for establishing tropical rainforest protection areas (e.g. national parks).

NAILIT!

Debt-for-nature swap

The lending organisation, such as the World Bank, will negotiate with the government of a country that owes it lots of money to agree to write-off some of the debt, as long as the country promises to legally protect a specified size of natural area. This benefits the country (reduced debt) and the world (saves biodiversity).

CHECKIT!

1 How does a tropical rainforest affect the local climate?

2 Explain why the soils in the tropical rainforest biome usually lack nutrients.

3 How long is the growing season in a tropical rainforest?

4 Give one reason why a country may wish to cut down its tropical rainforests.

5 Explain why the rate of deforestation in an area of tropical rainforest may vary over time.

6 Suggest the advantages and disadvantages of using the 'debt-for-nature swap' sustainable strategy.

Hot deserts

Characteristics of hot deserts
Physical characteristics

- Hot deserts have an extreme climate compared with many others on Earth. They are very dry (less than 250 mm of rainfall per year) and without much water. Life is very difficult for the natural environment and people. Daytime temperatures are very high – in some places over 50°C – but much cooler at night (diurnal temperature range). At night there are no clouds to trap heat near the ground so the hot air rises.

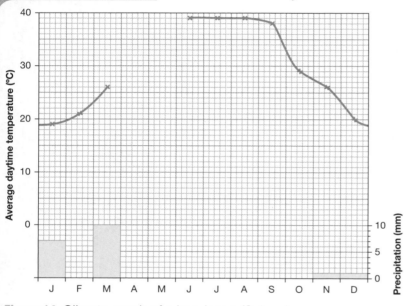

Figure 14 Climate graph of a hot desert (Sahara)

- The ecosystem is not very biodiverse, with few plants and animals, due to the lack of water and poor soils.

- Deserts are often thought of as a sea of sand with dunes, but many desert areas are rocky or consist of salt flats, as well as having sand dunes.

- Deserts may increase in size if surrounding areas become drier or are mismanaged by people – a process called desertification.

 Graphical skills

Climate graph (line and bar graph)

Make a copy of Figure 14.

1 Using the data given below, accurately plot the missing points for temperature to complete the line graph.

2 Using the data given below, accurately plot the missing bars for rainfall to complete the bar graph.

Month	Temperature (°C)	Precipitation (mm)
April	32	7
May	36	1

Interdependence

As in all ecosystems, the different parts are linked together and depend on each other, especially in a food web where there are many more plants than animals.

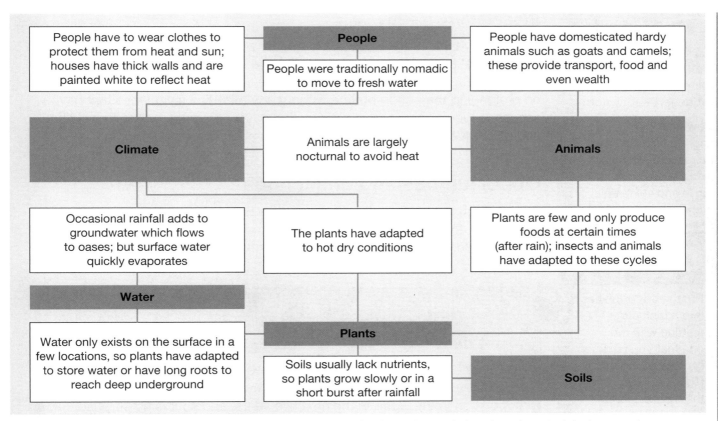

Figure 15 Links between climate, water, soils, plants, animals and people in a hot desert global ecosystem

Indigenous people, such as nomadic tribes, have adapted to the climatic conditions and natural patterns. Crops are difficult to grow, so nomadic tribes keep animals such as goats and camels and move them from one water source to the next. These animals provide food, clothing and products to trade.

Adaptation of plants and animals

Adaptations of plants:

- Long, deep roots (e.g. mesquite tree) or wide shallow roots (e.g. cactus) to gather as much water as possible.

- Able to store water within the plant for future use.

- Growing cycle linked to the occasional heavy rain with rapid growth, flowering and spreading of seeds.

- Thick outer layers with waxy skin to reduce water loss.

- Cactus spines are modified leaves that block sunlight to reduce overheating and water loss; they also provide some protection from being eaten.

Adaptations of animals:

- Camels store fat in their humps so that they can keep going without food and water.

- Many animals only become active at night in the cooler temperatures.

- Some rely on getting water from the food they eat.

- Some beetles collect water when it condenses on their bodies in the early morning.

SNAPIT!

Interdependence in a hot desert

Snap an image of Figure 15. Use it to learn three links within the hot desert ecosystem.

NAILIT!

Desert food web

In a desert, the most common living things are plants. However, these are fewer in number than in any other large-scale global ecosystem (biome), so there are few consumers at the higher levels of the food web.

Biodiversity issues

Deserts have low biodiversity. Although the plants and animals have adapted to the climatic conditions, they are still vulnerable to change – especially climate change caused by emissions of greenhouse gases, which appears to be prolonging the period of time without any rain. The few places that have access to water under the ground, called oases, are focal points for wildlife and also have pressures from increased numbers of settled people.

Opportunities and challenges

Even though the hot and dry desert climate presents human survival challenges, there may be opportunities linked to its geographical location or the presence of resources, such as oil or solar energy.

Case study

A hot desert: the Thar Desert

The Thar Desert is located on the border between India and Pakistan in southern Asia. India is set to become the country with the largest population in the world in the next five years, and so it is no surprise that there are many people trying to live in this desert.

Development opportunities

- There are valuable minerals that can be used by industries or in construction, such as gypsum and marble.
- There are energy resources such as lignite coal and oil at Barmer, which help increase energy security for India.
- There are opportunities to use renewable energy resources such as solar power at Bhaleri and wind power at Jaisalmer, which improve local quality of life.
- The landscape has attracted tourists, especially at the time of the Desert Festival.
- With irrigation water, for example, from the Indira Gandhi Canal, commercial farm crops can be grown such as wheat.

Development challenges

- The extreme heat in summer makes working outside very difficult.
- There are high evaporation rates from irrigation canals and farmland, and salts may be deposited in fields, killing plants.
- Water supplies are limited and so create problems for the growing number of people.
- Access through the desert region is tricky as roads are difficult to build and maintain due to the heat and sand movement.

DO IT!

Vulnerability of the hot desert ecosystem

Draw a revision diagram summarising the reasons why a hot desert ecosystem is fragile and vulnerable to change.

DO IT!

1 Put the opportunities into a rank order, starting with the one that you think is most important. Give two reasons for the most important opportunity being first and two reasons for your least important being last.

2 Put the challenges into a rank order, starting with the one that you think is the greatest challenge. Give two reasons for the biggest challenge being first and two reasons for your least important being last.

NAILIT!

Hot desert locations

The Thar Desert is at the same latitude as the Arabian Desert, Sahara Desert and Chihuahuan Desert. Remember that this is because the descending part of the Hadley convection cell is found at this latitude all around the world.

Desertification

Desertification means turning **semi-arid** areas (or drylands) into desert (see Figure 16).

Causes

Climate change

1. Natural climate change may be responsible for slowly turning semi-arid areas into desert.

2. The average temperature on Earth has increased by about 1°C since the industrial revolution started releasing carbon dioxide and other greenhouse gases into the atmosphere.

3. In the last 40 years, climate change appears to be changing the rainfall patterns in desert and semi-arid regions of the world so that droughts are getting longer, sometimes lasting for several years.

4. When any rain occurs it tends to be very heavy and so does not have time to soak into the ground, quickly running away over the ground surface. In this way, semi-arid areas are becoming even drier.

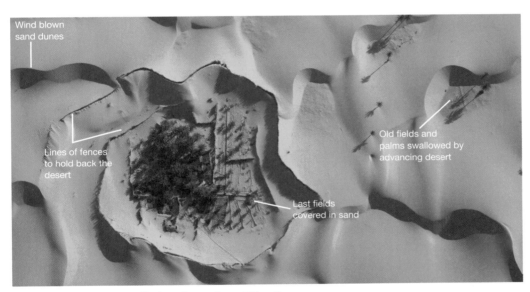

Wind blown sand dunes

Lines of fences to hold back the desert

Old fields and palms swallowed by advancing desert

Last fields covered in sand

Figure 16 An oasis suffering from desertification

Population growth

1. The population of the world has increased dramatically and this includes semi-arid areas. More people means that more food and water are needed, putting pressure on the limited water supplies and forcing the land to be used more intensively for crops or to graze animals. The land can be overused so that nothing grows and the desert takes over.

2. The traditional nomadic way of life, where relatively small groups of people move between oases or seasonal wetter areas, has rapidly declined due to enforcement of country boundaries and the permanent settling of the small areas capable of supporting farming. Therefore, more people are found in an area that can barely support them, which puts pressure on water supplies; taking too much out of the ground dries up wells and oases, or people grow crops to sell and these may take all the nutrients out of the soil so that it becomes useless and turns into desert.

Causes of desertification

Create a diagram to show how the possible causes of desertification are often linked.

Climate change and population growth

Climate change and population growth affect many aspects of geography. While population growth is slowing, and so it is possible that these pressures may decline later this century, pressures from climate change (natural and human made) are predicted to last well into the next century and perhaps beyond.

3 As some areas are changed into desert, people are forced to migrate to a nearby area which increases population pressures in these areas and makes the situation worse.

Removal of fuel wood

With a greater number of poorer people in a desert environment there is an increased demand for a fuel source to use for cooking, i.e. wood. There may be scattered trees and, while deadwood is gathered, some trees are also cut down. When this happens, the roots no longer hold the soil together and humus and moisture are reduced. The drier unprotected soil is more likely to be eroded by the wind, then nothing can grow and the desert takes over.

Overgrazing

As more people are forced to live in settlements in semi-arid areas, their livestock – often used instead of money – are concentrated in a few fields with little grass (see Figure 15 on page 47). Once the grass has been eaten, the roots no longer hold the soil together, so it can be blown away by the wind, enlarging the desert.

Over-cultivation and soil erosion

Overgrazing and population pressures have been explained above. Both of these contribute to over-cultivation – which is where people use the land too much, beyond the level at which it can recover after use. Soil is degraded to a point where nothing will grow in it and then the wind and occasional heavy rain can carry it away (erosion).

Strategies to reduce risk

Water and soil management

1 More efficient ways of getting the water out of the ground and irrigation methods can make the water last longer. Using **appropriate technology** (also known as **intermediate technology**) is relatively cheap and easy to repair. This includes lining wells with concrete, installing hand pumps such as Afridev or PlayPump, or using rain barrels to collect and store the occasional rain.

2 Building terraces on slopes to create flat land, which reduces run-off, or building small rock dams in gullies or wadis to trap water, or building rock walls (bunds) across fields to trap water and soil.

3 Soil can be conserved by not overusing it and protecting it from erosion by building wooden windbreaks or planting rows of trees and bushes. When cultivating, creating furrows across slopes rather than up and down them helps to trap any rainfall and gives it time to soak into the soil.

Tree planting

There are some types of tree or bush that have adapted more to arid environments. These can be planted as much as possible to shelter the ground from the wind, hold the soil together with their roots, add humus to the soil and create shade that reduces evaporation of water from the soil (see Figure 17).

DO IT!

Causes of desertification

Create a podcast or blog discussing the causes of desertification. Explain which are the most difficult to find solutions for.

NAILIT!

Appropriate technology

Appropriate technology is sometimes also known as intermediate technology. This is a level of technology that is suitable for low income countries (LICs) and also newly emerging economies (NEEs) because it is less expensive, avoids using expensive energy sources, is easy to understand and easy to repair. It is in between low-tech and high-tech.

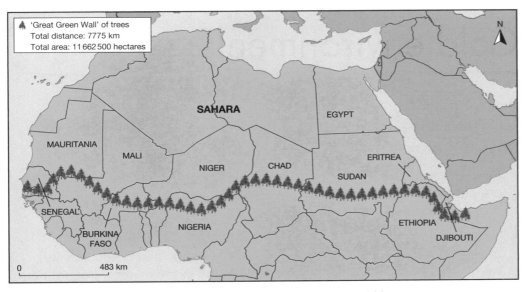

Figure 17 The Green Wall being created across the Sahel in Africa

Use of appropriate technology

As shown in the strategies above, using technology appropriate to the local people can be important, because they need to be able to survive without dependence on external help. This technology is sometimes described as 'intermediate' because it is in between 'low-tech' (rope and bucket for a well) and 'high-tech' (electronically controlled diesel water pump). It is fairly cheap and easy to understand so that local people can operate, repair and maintain it and it does not use expensive energy sources.

1 Using small metal cookers that require a smaller quantity of wood, or solar cookers, rather than using open fires, reduces the amount of vegetation cut down.

2 Building dams or bunds out of rocks that are a local material is cheaper than concrete dams and means they are more easily repaired.

Appropriate technology

Make a list of all the appropriate technologies that could be used by people to help them reduce the risk of desertification.

1 Give two characteristics of a hot desert climate.

2 Explain why a hot desert ecosystem can be considered to be fragile.

3 a Name and locate a hot desert ecosystem that you have studied.

b Describe three development opportunities that exist in this desert.

c Explain three development challenges that exist in this desert.

4 Explain how overgrazing may lead to desertification.

5 Suggest whether climate change or population growth is the most important cause of desertification. Explain your choice.

6 a Suggest why it is important to try to prevent further desertification taking place.

b Choose one strategy for reducing desertification and explain how it would help.

Cold environments

NAILIT!

The very cold temperatures have a major influence on what polar and tundra areas are like. Even though there is only a little snow each year, it is so cold that the snow does not melt but adds a thin layer which squashes (compresses) those underneath, so that eventually ice is formed. Where there is not much snow, the ground itself freezes solid for most of the year (permafrost).

Characteristics of cold environments

Physical characteristics

- Polar environments are found closest to the North and South Poles and are characterised by ice sheets and glaciers with occasional exposed mountain tops.

- Temperatures in Antarctica (cold desert) may drop below −80°C and there is very little precipitation. The central ice sheets in Greenland and Antarctica support no life and life is only found in the oceans around them or on the edges of the land masses.

- In tundra environments away from the poles very low temperatures are experienced, often well below −50°C with low precipitation (highest in summer), still making it difficult for plants and animals to live on land. There is also a lack of sunlight for over a month each winter. However, in the brief summer, temperatures may reach double figures (over 10°C) and this is when wildlife is most active.

- The ground is frozen solid for a large part of the year (permafrost) and only a thin surface layer thaws out in the short summer (active layer). The freezing and thawing causes movement in the active layer, which creates patterns within the ground as the process sorts soil and rock particles by size. On slopes, a slow movement takes place called solifluction.

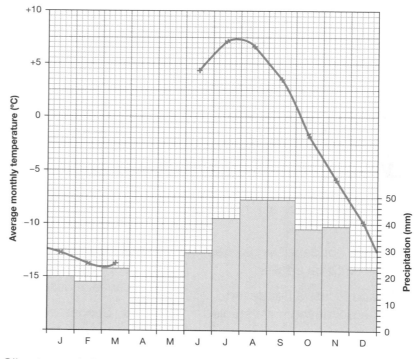

Figure 18 Climate graph for a tundra ecosystem (Greenland)

Graphical skills

Climate graph (line and bar graph)

Make a copy of Figure 18.

1 Using the data given below, accurately plot the missing points for temperature to complete the line graph.

2 Using the data given below, accurately plot the missing bars for rainfall to complete the bar graph.

Month	Temperature (°C)	Precipitation (mm)
April	–7	28
May	0.2	19

Interdependence

As in all ecosystems, the different parts are linked together and depend on each other, especially in a food web with lots of lichen and insect species.

Indigenous people, such as the Inuit peoples, have adapted to the cold climatic conditions and natural patterns. As it is not possible to grow crops, the hunting of animals, such as caribou or seals, and catching fish are important to their survival, providing food, tools, clothing and shelter.

SNAPIT!

Interdependence in a cold environment

Snap an image of Figure 19. Use it to learn three links within the tundra ecosystem.

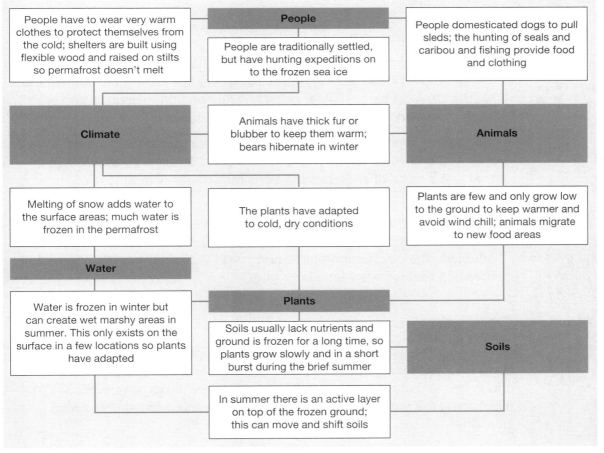

Figure 19 Links between climate, water, soils, plants, animals and people in a tundra ecosystem

NAILIT!

Animal migrations

As food supplies change during the seasons of the year some animals, like the caribou, migrate to find food, while others hibernate in winter after putting on as much weight as possible in the short summer, e.g. bears.

STRETCHIT!

Plants and animals

Know some names of plants and animals to add detail to your answers: plants such as Arctic poppy, Arctic willow, Arctic lichen, caribou moss and tufted saxifrage; animals such as Arctic fox, Arctic hare, Arctic wolf, snowy owl, brown bear and polar bear. Can you link some of these in a food chain?

NAILIT!

Cold environments and food webs

In a cold environment, the number of plants and animals is very low compared with other large-scale global ecosystems (biomes) because it is so cold. In polar areas, life is often based in and around the sea because the sea is warmer than the land in winter.

Adaptation of plants and animals

Adaptation of plants:

- Plants grow rapidly in the waterlogged ground of the tundra in the summer.
- Plants grow close to the ground to find shelter from the bitterly cold winds.

Adaptation of animals:

- White camouflage is used by grazing animals (herbivores such as the Arctic hare) or lower carnivores (e.g. the harp seal) in the tundra so that they cannot be so easily spotted against snow and ice by secondary consumers (e.g. the Arctic wolf). Polar bears hunting in polar areas are white so that seals cannot so easily see them.
- Caribou have wide hooves to help them move over waterlogged ground in the tundra and two layers of fur to keep them warm. Bears are omnivorous, eating fish and berries, for example, to be flexible according to the seasonal food supply.

Biodiversity issues

- Polar food webs are simpler than others and are based on the sea, such as the Southern Ocean around Antarctica where krill are a focal point in the food web. A problem in polar oceans is human commercial fishing activity, such as whaling, and even plans to harvest krill.
- Tundra food webs on land are a little more complicated, but are fragile due to the difficulty of surviving in the very cold climate, which means that just a small change in conditions may cause a large loss of biodiversity.
- The lack of biodiversity makes both polar and tundra cold environments fragile; if just a small part is damaged, then the whole food web can be affected. The cold temperatures also mean that plant growth is very slow, meaning that recovery time is lengthy.

Opportunities and challenges

Cold environments provide opportunities for people, such as the extraction of oil, scientific research and tourism. These environments also present challenges for human survival and development of the regions, while maintaining the unspoilt conditions in Antarctica and more remote Arctic locations.

Case study

A cold environment: Svalbard

Svalbard (Norwegian territory) is a small group of islands in the Arctic Ocean, located north of Norway and east of Greenland at a latitude around 80°N. There is a small population of just under 3000 people, most of whom live in the town of Longyearbyen.

Development opportunities

- Coal reserves exist and provide jobs. They are used to run a thermal power station for the islands' electricity, as well as providing exports.
- Renewable energy could be developed by using geothermal sources or capturing and recycling the carbon emissions from the coal-fired thermal power station.
- Svalbard is surrounded by seas with rich fishing grounds, such as the Barents Sea. Cod and other fish are caught by trawlers and exported.
- Tourism has grown with cruise ships arriving from mainland Norway and further afield, as people want to see the natural environment and participate in activities.

Development challenges

- Coal is a fossil fuel and its use in power stations produces carbon dioxide emissions.
- Too many tourists could damage the natural environment that they have come to see.
- It is dark for four months of the year and very cold. People always have to wear very warm clothes.
- Buildings and infrastructure (roads, water and sewerage) have to be constructed with the very cold weather in mind, so all of these are raised above the ground so that they don't melt the permafrost and collapse or break.
- Transport to and from, and between, the islands is very difficult.
- Overfishing or pollution of fish stocks is a constant concern.

Vulnerability of cold environments

Draw a revision diagram summarising the reasons why a tundra or polar ecosystem is fragile and vulnerable to change.

Cold environment challenges

People living and working in cold environments not only face challenges due to the extreme climate, but these areas are usually very remote and essentially cut off from the rest of the world.

Polar and tundra opportunities and challenges

1 Put the opportunities into a rank order, starting with the one that you think is the most important. Give two reasons for the most important opportunity being first and two reasons for your least important being last.

2 Put the challenges into a rank order, starting with the one that you think is the greatest challenge. Give two reasons for the biggest challenge being first and two reasons for your least important being last.

Risks from economic development

Value as wilderness

There are few places left on Earth that have not been used by people, and even these are likely to have been affected by climate change brought about by emissions of greenhouse gases. Therefore, cold environments are worth protecting (e.g. the Antarctic Treaty and its protocols; see page 57).

- Some tundra areas in northern Canada and the white continent of Antarctica are **wilderness** areas where humans do not dominate the landscape. In these areas, natural processes and systems take place without interference and are therefore wild.

- Cold environments have genetic diversity for scientific study, which may yield future benefits for people (e.g. medicines, space exploration).

- Antarctica is valuable as the thick ice has trapped air and particles from a very long time ago. Scientists can study these and learn a lot.

- Antarctica regulates the temperature of the whole world by cooling it.

- Pristine cold environments (see Figure 20) allow comparisons with areas that have been changed, reminding people of the importance of the beauty of nature and the cultures of indigenous peoples.

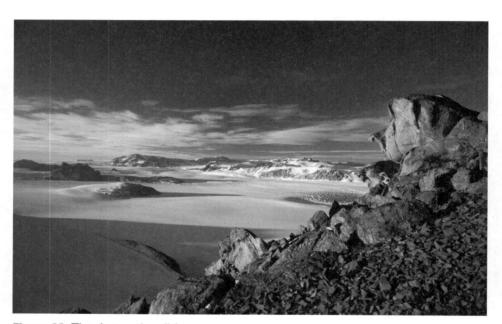

Figure 20 The Antarctic wilderness area

Strategies to achieve balance between economic development and conservation

Use of technology

- Oil extraction in Alaska needs to be done with great care. In particular, the oil, which is warm, has to be transported by pipeline from northern Alaska to its southern coast through the tundra area. This has been made possible by raising the pipeline up on stilts and adding radiators to take the heat from the oil away into the air, which stops the warm oil melting the permafrost.

STRETCH IT!

Wilderness in 2016

Wilderness areas are disappearing. For example, 3.4 million km² has been lost in the last 25 years. About 20 per cent of the world's land area is still classified as wilderness – that's 30 million km². So at the rate of disappearance over the last decade it could all be gone within 200 years.

DO IT!

Wilderness value

Summarise the importance of wilderness areas for the world.

- The expansion of ICT allows indigenous communities access to education and health care and to preserve culture in a knowledge network. Internet companies have located in tundra environments to save costs on cooling the computer equipment.

Role of governments

- Governments have control over their territory and, therefore, have the ability to set aside areas just for nature – such as the Arctic National Wildlife Reserve in Alaska – or to only issue a limited number of oil exploration licences. In this way, they can balance the need to make money and find resources that can help develop the country with the need to protect nature from serious damage.

- Governments can ensure that they follow all international treaties, protocols and laws; for example, the UK passed the Antarctic Act in 2013.

Figure 21 Antarctica with flags of treaty countries

International agreements

- Antarctica is protected by a treaty that has been in force since 1961. All the countries (see Figure 21) that have claimed a slice of the continent have agreed not to pursue their claims and the exploitation of minerals is currently banned. In 1998, an Environmental Protection protocol came into force to further protect nature from human activities, such as the small tourist business. However, what is taking place is international research into current and past climates, the condition of the atmosphere, wildlife patterns and chemicals used by marine organisms to survive cold conditions.

- Whaling has been banned since 1986 but, while most countries abide by this law, some do not.

- An Arctic Council has been set up to include all those who live on the land around the Arctic Ocean, with the idea of jointly working towards sustainable development.

- Climate change agreements, such as Paris 2015, aim to reduce greenhouse gas emissions and this may limit the amount of melting ice and permafrost.

DO IT!

Four examples of international agreements have been listed. Create a mnemonic to help you remember them.

NAIL IT!

International agreements

When countries sign a treaty, they agree to be bound by its directives. For example, the Antarctic Treaty has ensured that the 'white continent' has not been exploited and its Environment Protocol has ensured that the amount of damage from scientific bases and tourism has been greatly reduced. However, some laws are ignored or deliberately misinterpreted – such as the ban on whaling, with some countries using the 'scientific research' clause to justify catching whales.

Role of conservation groups

There is a lot of concern that some countries continue with whaling. Whaling is banned commercially by international law, but some is allowed for scientific research. Some countries are exploiting this and conservation pressure groups such as Greenpeace have conducted campaigns to actively disrupt whaling. Greenpeace has also campaigned against oil exploration in the Arctic and has suggested that there should be an Arctic global sanctuary similar to that in the Antarctic.

CHECK IT!

1 Give one characteristic of the polar environment.

2 **a** Name an area of the world with a polar ecosystem.

   b Name an area of the world with a tundra ecosystem.

3 **a** Describe two features of a tundra landscape.

   b Describe two features of a tundra climate.

4 Explain what is meant by the term 'pristine wilderness'.

5 Explain why Antarctica is important to the world.

6 Explain the role of international agreements in helping to protect cold environments.

7 Suggest whether governments or conservation groups have the most important role in protecting cold environments. Explain your choice.

8 **a** Suggest why it is important to try to conserve cold environment wilderness areas.

   b Choose one strategy for conserving a cold environment wilderness area and explain how it would help.

Ecosystems

1 **a** State one important natural influence on a small-scale ecosystem.

 b State one important natural influence on a large-scale global ecosystem.

2 For a small-scale UK ecosystem that you have studied:

 a State two main characteristics of the natural ecosystem.

 b Explain one human pressure that may disrupt the links within the ecosystem.

 c Explain one natural pressure that changes the links within the ecosystem.

Tropical rainforests

1 **a** Describe the structure of a tropical rainforest.

 b Explain why there is little vegetation at the ground level in a tropical rainforest.

2 **a** Where do most creatures live in the tropical rainforest structure?

 b Give one reason why most creatures live in this part of the structure.

3 Explain why there is a very wide range of adaptations of plants and animals to the environmental conditions in a tropical rainforest.

4 Outline one issue caused by a loss of biodiversity in a tropical rainforest.

5 **a** State one likely local impact of cutting down part of a tropical rainforest.

 b State one likely global impact of cutting down the tropical rainforest.

6 Compare the impact of indigenous people and people with strong economic motives on the tropical rainforest.

7 For a tropical rainforest that you have studied:

 a Explain four causes of deforestation within this tropical rainforest.

 b Describe the impact on the natural environment of the deforestation in this tropical rainforest.

 c Describe the impact on people within the country where this deforestation is taking place.

8 **a** Why do so many people think that it is important to protect the tropical rainforest global ecosystem?

 b Choose one sustainable strategy for looking after the tropical rainforest and explain how this strategy would help.

Hot deserts

1 **a** What is meant by the term 'diurnal temperature range'?

 b What is the diurnal temperature range like in a hot desert?

2 **a** Explain two natural links within a hot desert ecosystem.

 b Explain one link between nature and people in a hot desert ecosystem.

3 **a** Explain why biodiversity in a hot desert ecosystem is much lower than in most other global ecosystems.

 b Give two reasons why hot desert ecosystems are vulnerable to change.

4 What is meant by the term 'semi-arid'?

Cold environments

1 **a** State two natural links within a cold environment ecosystem.

 b Explain one link between nature and people in a cold environment ecosystem.

2 **a** Explain why biodiversity in a polar or tundra ecosystem is much lower than in other global ecosystems.

 b Give two reasons why cold environment ecosystems are vulnerable to change.

3 For a cold environment ecosystem that you have studied:

 a Explain why people want to live and work in this cold environment.

 b Explain the pressures and challenges faced by people in this cold environment.

Physical landscapes in the UK

UK physical landscapes

THEEXAM!

- This section is tested in Paper 1 Section C.
- You must know UK physical landscapes and **two** of coastal landscapes in the UK, river landscapes in the UK and glacial landscapes in the UK.

Relief describes the physical shape of the land, including height above sea level, steepness of slopes and shape and orientation of landscape features.

The relief of the land is determined by its geology. Harder more resistant rocks, such as granite, will form mountain ranges, whereas softer more easily eroded rocks, such as clay, will form more low-lying landscapes.

DO IT!

Study Figure 1.

1 Describe the pattern of upland areas in the UK.

2 Describe the pattern of lowland areas in the UK.

SNAP IT!

Find a relief map of the UK, draw it and take a picture of it. Use it to help you remember the UK's:

- areas of uplands and lowlands
- extensive river system.

Figure 1 Physical map of the UK

NAIL IT!

Make sure you can explain how different rock types have determined the UK's landscape.

CHECK IT!

1 Give a correct definition of the term 'relief'.

2 Describe the course of the River Thames from source to mouth.

3 Explain how different rock types affect the landscape.

Coastal landscapes in the UK

Physical processes

Wave types

Waves are formed by the wind blowing over the sea. The size of a wave is affected by the:

- speed of the wind
- length of time the wind has been blowing
- distance the wind blows across the water – the **fetch**.

Figure 2 When waves reach the coast

There are two main types of waves: **constructive waves** and **destructive waves**.

Constructive waves:

- gently sloping
- strong swash
- weak backwash
- long wavelength
- deposit large amounts of sediment.

Destructive waves:

- steep wave
- weak swash
- strong backwash
- short wavelength
- remove large amounts of sediment.

DOIT!

Wave characteristics

Draw an annotated diagram to show the difference between a constructive wave and a destructive wave.

SNAPIT!

Wave characteristics

Snap an image of Figures 3 and 4. Use them to remind yourself of the differences between a constructive and destructive wave. Think about why they are different.

Figure 3 Constructive wave

DOIT!

Chemical weathering

Research two types of rock affected by chemical weathering.

Mechanical weathering

Draw a set of annotated sketches to show the process of freeze-thaw weathering.

Biological weathering

Research an image from the internet to show an example of biological weathering.

NAILIT!

Coastal processes

Make sure that you understand the different forms of weathering and how they can influence cliff collapse. Freeze-thaw weathering weakening a cliff face over time may result in rock fall.

DOIT!

Mass movement

Draw a revision poster to show the different types of mass movement. Try to include annotated sketches to help you understand the different types of mass movement.

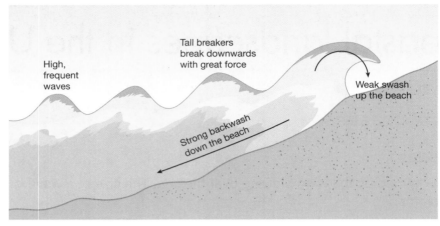

Figure 4 Destructive wave

Coastal processes

Weathering processes

There are three types of weathering:

1. **Chemical weathering**: caused by chemical changes. For example, slightly acidic rainwater can dissolve certain rock types.

Figure 5 Freeze-thaw weathering

2. **Mechanical weathering**: causing the break-up of rocks. For example, freeze-thaw weathering.

3. **Biological weathering**: caused by the action of plants (flora) and animals (fauna). For example, plant roots growing into cracks in the rock.

Mass movement

Mass movement is the downward movement of surface material, for example, rock or soil under the influence of gravity. There are several main types of mass movement at the coastline:

- **Rock fall**: rocks break away from the cliff.

- **Landslides**: blocks of rock slide down the cliff face.

- **Mudflows**: weak rock and saturated soil flow down a slope.

- **Rotational slip**: a slump of saturated soil and weak rocks move along a curved surface.

Figure 6 Mass movement

Coastal marine processes

Coastal erosion

Erosion is the removal of sediment and rocks. There are three types of coastal erosion:

1. **Hydraulic action**: the power of the waves smashes into a cliff. The trapped air is forced into holes and cracks in the rock again and again with explosive release, eventually causing it to break apart.

Figure 7 Coastal transportation

2. **Abrasion/corrasion**: the wearing away of the cliffs by sand, shingle and boulders being hurled against them by waves.

3. **Attrition**: rocks in the sea knock against each other, causing them to become smaller and rounder.

Transportation

There are four ways that sediment can be transported:

1. Solution – dissolved minerals in the water.

2. Suspension – small particles that are carried with the water.

3. Saltation – particles that are too heavy to be suspended are bounced along the sea floor.

4. Traction – large particles that are rolled along the sea floor.

Longshore drift is the process by which sediment is moved along the beach.

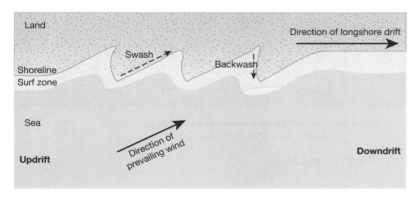

Figure 8 Longshore drift

Deposition

When water loses its energy, sediment can no longer be carried or transported and so it is deposited. Several conditions cause deposition:

- Sheltered bays where constructive waves are dominant.

- Areas where there are large expanses of flat beach.

- Where material is trapped behind a spit.

- Where structures, such as groynes, trap sediment.

NAILIT!

Coastal marine processes

Remember the different types of erosion, transportation and deposition. You will need to know these to understand why we get different landforms along the UK coastline.

DOIT!

Coastal erosion

Create a revision card for each type of coastal erosion. Include an annotated sketch to explain how each works.

SNAPIT!

Longshore drift

Take a picture of Figure 8. Use it to remind you of how longshore drift happens. Remember that swash always moves in the direction of the prevailing wind and backwash is always at 90 degrees to the coastline.

Coastal landforms

Factors affecting coastal landforms

- **Geological structure**: this is the way that rocks are folded or tilted. This is an important factor affecting the shape of cliffs.

- **Rock type**: harder, more resistant rocks such as granite and limestone are more resistant to erosion, whereas softer rocks such as clay and sandstone are more easily eroded.

Coastal erosion landforms

Headlands and bays

- Soft rock and hard rock are eroded at different rates.
- Weaker bands of rock erode more easily to form bays.
- More resistant bands of rock erode more slowly and stick out to form headlands.

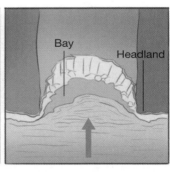

Coastline changes over time

Figure 9 Formation of headlands and bays

Cliffs and wave cut platforms

1. Weathering processes weaken the rock face.
2. Marine erosion processes, such as hydraulic action and abrasion erode the base of the cliff.
3. The base of the cliff starts to wear away and destructive waves remove this material. A wave cut notch is formed, which results in the cliff above being unsupported.
4. The notch will become bigger and bigger due to erosional processes, resulting in the cliff above eventually breaking away.
5. This process continues to happen over time, leaving behind the former base of the cliff as a wave cut platform (see Figure 10).

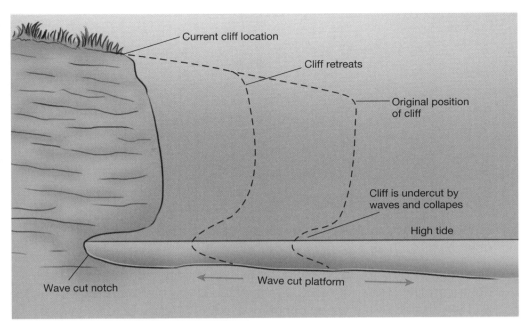

Figure 10 Cliff retreat and formation of a wave cut platform

DO IT!

Cliffs and wave cut platforms

Study Figure 10. Create a flow diagram to show the process of how a wave cut platform is created. You may want to include diagrams to help you remember the formation.

Caves, arches and stacks

Figure 11 Formation of caves, arches and stacks

1 There is a large crack/line of weakness in the headland.

2 The crack is enlarged by the process of hydraulic action and abrasion to form a notch.

3 Waves make the notch larger over time, forming a cave.

4 Two caves form back-to-back and will eventually cut through the headland to form an arch.

5 Weathering processes weaken the top of the cliff. It will become too heavy and collapse, leaving a stack.

6 The stack is eroded over time and will collapse and leave a stump.

Coastal deposition landforms

Beaches

A beach is the deposit of sand and shingle; materials that are found in sheltered bays. Beaches are formed due to constructive waves, which have a strong swash, carrying the material up the beach and depositing it.

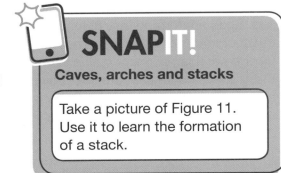

SNAPIT!

Caves, arches and stacks

Take a picture of Figure 11. Use it to learn the formation of a stack.

Sand dunes

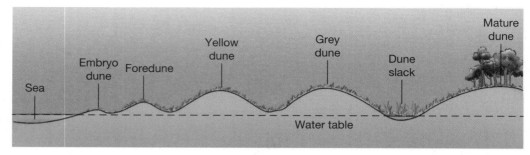

Figure 12 Development of sand dunes

1. Embryo dunes are closest to the sea and form around deposited obstacles, such as pieces of wood.

2. The embryo dune builds up and the growth of vegetation stabilises the dune, forming a foredune and a yellow dune. Marram grass grows quickly and is adapted to the windy conditions. It has long roots to find water and these help to stabilise the dune by binding the sand together.

3. As vegetation dies it adds organic matter to the sand, which makes it more fertile. This allows a greater range of plants to colonise the back dunes.

4. Winds can form depressions between lines of dunes called dune slacks.

Spits and bars

Figure 13 Formation of a spit

A spit is a long finger of sand or shingle that sticks out from the land into the sea. Spits are formed due to longshore drift, where the coastline changes direction and bends sharply, resulting in sediment being deposited out to sea. Over time, this sediment builds up, forming an extension from the land. As the process continues, the spit grows further out into the sea. If a spit grows across a bay, trapping a freshwater lake behind it, this then forms a bar. In the sheltered area behind the spit, a saltmarsh may form.

STRETCH IT!

Sand dunes

Research the reason why sand dunes only form along certain coastlines.

DO IT!

Formation of a spit

1 Using Figure 13, create a flow diagram explaining the formation of a spit.

2 What is the difference between a spit and a bar?

Management strategies to protect coastlines from physical processes

There are three different types of coastal management:

1 **Hard engineering**: the use of human-made structures to protect the coastline.

2 **Soft engineering**: methods that work with natural processes to protect the coastline.

3 **Managed retreat**: the controlled retreat of the coastline.

Hard engineering

Sea walls

A sea wall is a concrete wall built at the base of a cliff or along sea fronts in towns. Many sea walls have a curved face or slope away from the sea to reflect the power of the waves back out to sea.

Advantages of sea walls	Disadvantages of sea walls
• effective at reflecting the power of the sea • can last for many years if well maintained • used as walkway or promenade • don't affect the movement of sediment	• can be visually unappealing • very expensive • can restrict access to the beach

Table 1 Advantages and disadvantages of sea walls

Rock armour

Rock armour is piles of large boulders arranged at the base of a cliff to reduce the power of waves.

Advantages of rock armour	Disadvantages of rock armour
• relatively cheap • structure is easy to build and maintain • lasts a long time if well maintained	• rocks used are taken from other coastlines or abroad • does not fit in with the local geology • visually unappealing

Table 2 Advantages and disadvantages of rock armour

Gabions

Gabions are wire cages filled with rocks and placed to support a cliff and to take the power of the waves.

Advantages of gabions	Disadvantages of gabions
• cheap to make and easy to construct • can improve the drainage of a cliff • over time vegetation will grow, helping them to blend into the coastline	• in the short term they look very unattractive • cages only last for 5–10 years before rusting

Table 3 Advantages and disadvantages of gabions

Groynes

Groynes are wood or rock structures built out to sea from the coastline. They trap sediment moved by longshore drift, making the beach larger. A wider beach will reduce the power of the waves.

Advantages of groynes	Disadvantages of groynes
• create a wider beach, which is good for tourists • not too expensive • can last for a long time if well maintained	• they interrupt longshore drift, which starves beaches further down the coastline of sediment • can be unattractive

Table 4 Advantages and disadvantages of groynes

DO IT!

Rank the four types of hard engineering listed on pages 67 and 68, from the most to the least effective. Explain your order.

Numerical skills

A local council wants to protect 2 km of coastline using a hard-engineering technique. Which of the following would be the most cost effective?

- Sea wall: £5000 per metre.
- Rock armour: £1000 per metre.
- Groynes: £5000 each (placed every 200 m).
- Gabions: £110 per metre.

Soft engineering

Beach nourishment

Beach nourishment involves the addition of sand or shingle to a beach to make it wider.

Advantages of beach nourishment	Disadvantages of beach nourishment
• quite cheap and easy to maintain • blends in with the beach that is already there • good for tourists	• will need constant maintenance as sediment is removed through longshore drift • sediment has to be sourced • sediment has to be transported to the beach

Table 5 Advantages and disadvantages of beach nourishment

Reprofiling

Reprofiling involves the reshaping of the beach, as the beach is often lowered in the winter by destructive waves.

Advantages of reprofiling	Disadvantages of reprofiling
• blends in with the beach, keeping a natural look	• major reprofiling works can be expensive • a steep beach may be unattractive to tourists

Table 6 Advantages and disadvantages of reprofiling

Dune regeneration

Sand dunes are natural buffers, but can be easily destroyed during storms. During dune regeneration, marram grass is planted to help stabilise the dunes and fences are put up to stop people trampling across the dunes and causing erosion.

Advantages of dune regeneration	Disadvantages of dune regeneration
• small impact on the natural environment • can control the public's access to natural ecosystems	• can be unattractive • need to be regularly maintained, especially after storms

Table 7 Advantages and disadvantages of dune regeneration

Managed retreat

Managed retreat is the policy of allowing an area of low-value land to be naturally eroded or flooded.

Advantages of managed retreat	Disadvantages of managed retreat
• takes pressure off the land further down the coastline • cheaper than continuing to maintain hard engineering defences • good for the environment, as it encourages natural habitats to develop	• people may have to be relocated, which affects communities and can be expensive in the short term • large areas of agricultural land can be lost

Table 8 Advantages and disadvantages of managed retreat

Create a poster showing the advantages and disadvantages of managed retreat.

Coastal management

Make sure that you understand the difference between hard and soft engineering techniques and why they may be used in different areas along the UK coastline.

Coastal management

Write an answer explaining why you think hard or soft engineering is more effective at managing the UK coastline.

Research which areas of coastline would be suitable for managed retreat. Name some examples.

Case study

A coastal management scheme in the UK: Lyme Regis

Reasons for management

- The town has been built on unstable cliffs.
- Powerful waves from the south means the coastline is eroding rapidly.
- The sea walls have already been breached many times.

The management strategy

- The Lyme Regis Improvement Scheme was set up in the 1990s.
- There were four phases to the scheme:

  Phase 1 (1990s): a new sea wall was constructed to the east of the River Lim.

  Phase 2 (2005–07): construction of new sea walls; creation of wide sandy beaches; and extension of rock armour along the coastline. This all cost approximately £22 million.

  Phase 3: plans to further prevent landslips and erosion were not undertaken.

  Phase 4 (2013–15): construction of a new sea wall in front of the old one and improvement of cliff drainage to provide better cliff stabilisation. This phase cost £20 million.

How successful has the scheme been?

Costs:
- Tourist numbers have increased, which has led to conflict with the locals.
- Some of the defences are unattractive.
- The new sea wall may affect processes, which could cause conflict further along the coastline.

Benefits:
- Increased tourist numbers due to new beaches.
- Winter storms have not breached the new defences.
- The harbour now has better protection.

DOIT!

Lyme Regis coastal management scheme

Plan an answer to summarise the potential conflict which could arise from the coastal management scheme at Lyme Regis or the coastal management scheme you have studied, if different.

CHECKIT!

1 Describe the differences between a constructive wave and a destructive wave.

2 Explain the link between weathering and mass movement.

3 Name two coastal landforms created by erosion.

4 Describe how sand dunes change over time.

5 Describe the costs and benefits of two types of hard engineering.

6 Explain on which type of coastline you would decide to use managed retreat.

River landscapes in the UK

Changes in rivers and their valleys

A **drainage basin** (also known as a catchment area) is the area of land drained by a river and its tributaries (see Figure 14).

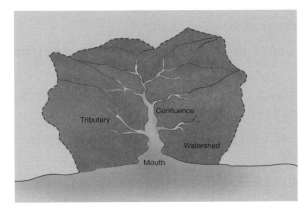

Figure 14 Drainage basin

A river is divided into three sections – the **upper**, **middle** and **lower courses**. Each course has a distinct valley shape, river velocity (speed) and bedload (sediment transported by the river) (see Figure 15 and Table 9).

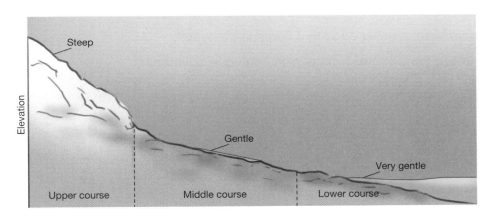

Figure 15 Long profile of a river

Upper course	Middle course	Lower course
• steep gradient • shallow depth • narrow, steep-sided channel • quite fast velocity • steep V-shaped valley • waterfalls and interlocking spurs	• gentle gradient • deeper • flatter channel with steep sides • fast velocity • U-shaped valley • meanders and floodplains	• very gentle gradient • very deep • flat channel with gentle sloping sides • very fast velocity • wide flat valley • meanders, ox-bow lakes, floodplains and levées

Table 9 Changes of a river from source to mouth

Changes in rivers

Using Figure 15, draw a sketch of the long profile of a river. Annotate it to show the different features of each course.

Fluvial processes

Rivers carry out three main processes as they move downstream:

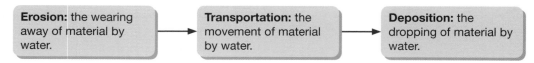

Erosion: the wearing away of material by water. → **Transportation:** the movement of material by water. → **Deposition:** the dropping of material by water.

Erosion

The ability of a river to wear away the land depends on its velocity (how fast it is travelling). Erosion can be **vertical** (downwards) or **lateral** (sideways).

There are four different types of erosion:

1. **Hydraulic action**: this is the force of the water hitting the river bank and river bed.

2. **Abrasion**: the load that is carried by the river scrapes along the bank and bed and wears them away.

3. **Attrition**: the stones carried by the river hit each other and break up into smaller pieces.

4. **Solution**: material is dissolved when the water travels over rocks, making the water acidic.

Transportation

A river transports, or carries, the load in four different ways:

1. **Traction**: large particles roll along the river bed.

2. **Saltation**: smaller particles bounce along the river bed.

3. **Suspension**: smaller sediment floats in the water.

4. **Solution**: the dissolved material is carried in solution.

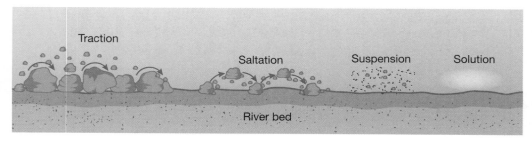

Figure 16 Types of transportation

Deposition

Deposition occurs when a river loses velocity and slows down. This is because it loses the energy to carry material.

- Large rocks get deposited in the upper course of the river, as they are too heavy to be carried very far.

- Smaller particles held in suspension are deposited when a river slows down, for example, on a river bend where there is greater friction.

- Most deposition takes place at the mouth of the river where the river loses energy when it meets the sea.

Fluvial landforms

Rivers have distinct landforms as you move down the long profile towards the sea.

Interlocking spurs

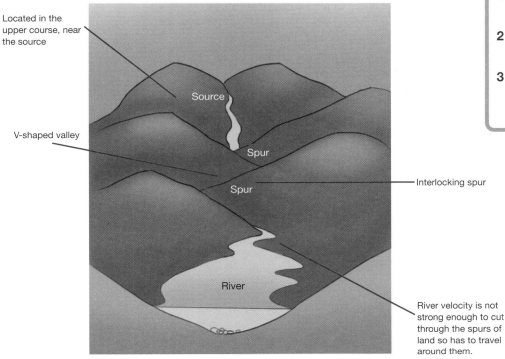

Located in the upper course, near the source

V-shaped valley

Source

Spur

Spur

Interlocking spur

River

River velocity is not strong enough to cut through the spurs of land so has to travel around them.

Figure 17 Interlocking spurs

Waterfall

Waterfalls occur where the river flows over two different rock types. The softer rock is less resistant and more easily eroded than the harder rock.

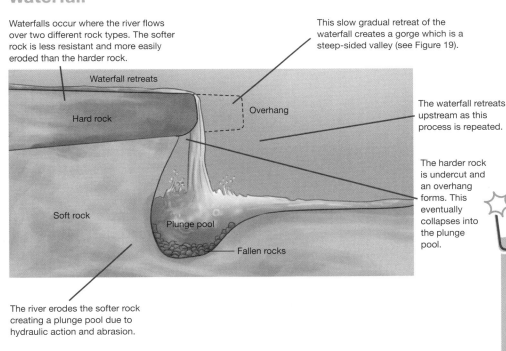

Waterfall retreats

Hard rock

Overhang

Soft rock

Plunge pool

Fallen rocks

This slow gradual retreat of the waterfall creates a gorge which is a steep-sided valley (see Figure 19).

The waterfall retreats upstream as this process is repeated.

The harder rock is undercut and an overhang forms. This eventually collapses into the plunge pool.

The river erodes the softer rock creating a plunge pool due to hydraulic action and abrasion.

Figure 18 Waterfall formation

DO IT!

Waterfall

Write a paragraph to explain how a waterfall is formed. Use all the following words:

- overhang
- retreat
- erosion
- gorge
- less-resistant rock
- plunge pool
- undercutting.

DO IT!

Meander

1 Write a paragraph to explain how a meander is formed. Describe the changes over time.

2 Draw a sequence of diagrams to show how a meander can lead to the formation of an ox-bow lake.

STRETCH IT!

Meander

Draw your own cross section through a meander. Annotate it to show the main features of a meander. You should include the river cliff, slip-off slope, areas of erosion and deposition and the thalweg.

Figure 19 Gorge

Meander

The thalweg is the fastest line of velocity in a river. It swings from the outside of one bend to the outside of the next bend.

This process of erosion on the outside of a bend and deposition on the inside of a bend causes the river to move across the valley floor.

- Lateral erosion
+ Deposition
➤ Fastest current
Land lost to the river
New land gained from the river

Inside of bend

The velocity is slower on the inside of a bend, causing deposition.

Outside of bend

The velocity is stronger on the outside of a bend causing erosion.

Figure 20 Meander

Ox-bow lake

The neck of a meander is gradually eroded and narrows.

During a period of flood, the neck of the meander is broken through and the river takes the course of least resistance to form a new straight channel.

ox-bow

Over time, the meanders start to move towards each other.

The meander is then cut off by deposition. This leaves behind an ox-bow lake.

Figure 21 Ox-bow lake

Floodplains and levées

A **floodplain** is a wide, flat area of land either side of a river.

Floodplains are formed:

- When a meander migrates from side to side causing the valley sides to erode and become wider. This creates a wide, flat valley floor.

- During a period of flood, silt is deposited on to the floodplain and builds up layers of fertile land.

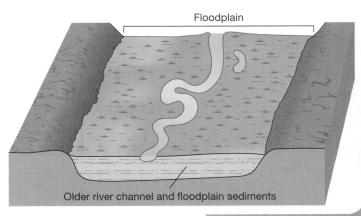

Figure 22 Floodplain

A **levée** is the raised river bed created by sediment from the river during a flood. Material is deposited at the sides of the river and builds up over time.

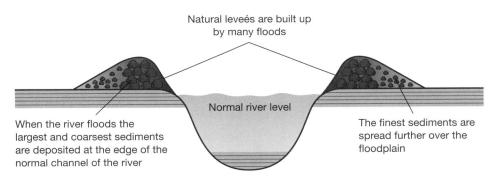

Figure 23 Levée

Flood risk and management

Factors that increase the risk of flooding

Flooding occurs when there is an increase in the volume of water and it overflows on to the land surrounding the river. There are a number of human and physical factors that can increase the risk of flooding (see Table 10).

Physical factors	Human factors
Precipitation: a sudden heavy downpour can lead to flash floods or long periods of steady rainfall can lead to flooding.	**Urbanisation**: creates impermeable surfaces, such as roads, and drains and sewers speed up the movement of water to the river.
Saturated soil: waterlogged ground will not allow the water to soak in.	**Deforestation**: the removal of trees means less interception of rainfall, so water gets to the river channels more quickly.
Geology: impermeable rock that doesn't allow water to soak in allows faster surface run-off to the river.	**Agriculture**: water travels more quickly along ploughed furrows and fields that are left unplanted.
Steep slopes: water will run off a steep slope in a mountainous area much more quickly.	

Table 10 Human and physical factors that can increase the risk of flooding

Hydrographs

Hydrographs show the amount or volume of water in a river (**discharge**) following a storm.

There are a number of factors that affect the shape of a hydrograph. A 'flashy' hydrograph has a short lag time and a high peak. This occurs in an area where there is:

- a small drainage basin
- impermeable rock
- urbanisation
- steep relief
- saturated soil
- heavy rainfall.

A low, flat hydrograph shows a long lag time and a low peak. This occurs in an area where there is:

- a large drainage basin
- permeable rock
- forest
- gentle relief
- dry soil
- light rainfall.

Graphical skills

For the hydrograph in Figure 24:

1 When was the peak rainfall?

2 When was the peak discharge?

3 Calculate the lag time.

NAIL IT!

Factors that increase the risk of flooding

Make sure you can describe the human and physical factors that can increase the risk of flooding. Explain how this can be demonstrated using a hydrograph.

STRETCH IT!

For each type of hydrograph, explain why these factors affect its shape:

- size of drainage basin
- land use
- rock type
- soil type
- amount of rainfall
- relief.

Figure 24 Flood hydrograph

Managing floods

Hard engineering

Hard engineering is the use of human-made structures to prevent or control flooding. These can be very expensive to install, but are often used to protect areas of high value, such as housing estates. Hard engineering schemes have a number of costs and benefits.

	Benefits	Costs
Dams and reservoirs	• regulate flow of water • store water • allow for slow release of rainfall • multi-functional (hydro-electric power (HEP), recreation, irrigation)	• very expensive • flood large areas of land permanently • people moved from their homes
Channel straightening	• speeds up the flow of water • protects the vulnerable area next to the channel	• increases flood risk downstream • silt can build up in concrete channels • concrete channels are unattractive and unnatural
Embankments	• raised river banks allow the river to hold more water during a flood • can be sustainable if dredged material is used • protect towns and cities from flooding	• high concrete walls can look unattractive
Flood relief channel	• new river channel bypasses the urban area • redirects the river away from a town during periods of heavy rain	• expensive

Table 11 The costs and benefits of hard engineering flood management schemes

Soft engineering

Soft engineering is when natural river processes are managed to reduce flood risk.

Afforestation

Afforestation is the planting of trees to increase interception and absorption of rainwater.

Wetlands and flood storage

Wetlands are areas of land that are allowed to deliberately flood and become flood storage areas.

Floodplain zoning

Floodplain zoning is where areas of land on the floodplain are divided up to allow certain types of land use. Land nearer the river is used for low-cost uses, such as parkland. High-cost uses, such as housing or industry, are kept further away from the river so they are protected from flooding.

River restoration

River restoration returns a river to its natural state and allows its natural features to slow down water flow, such as meanders and wetlands.

Managing floods

1 Describe the costs and benefits of two hard engineering flood management schemes.

2 Describe the costs and benefits of two soft engineering flood management schemes.

Soft engineering

Suggest why some people might prefer soft engineering schemes to hard engineering schemes.

Case study

A flood relief scheme: the Jubilee River flood relief channel

Why was the scheme required?

- The scheme was required to protect Windsor and Eton in southern England when the River Thames floods. This is a high-value area including Windsor Castle, Eton public school and a residential area.

What was the management strategy?

- The Jubilee River is a flood relief channel that was created to divert water from the River Thames when the water discharge is high and to prevent the Thames from overflowing its banks.
- The Jubilee River is 11.7 km long and 50 m wide. It cost the Environment Agency £110 million to build.
- It was created to look like a natural river, with meanders, reed beds and a nature reserve.

Figure 25 Jubilee River flood relief channel

What are the social issues?

- It protects the high-value areas of Windsor and Eton, but has increased problems in Old Windsor and Wraysbury areas downstream.

What are the economic issues?

- It is the most expensive flood relief scheme in the UK.
- Maintenance costs of the channel and weirs are high, as they can become damaged during floods.
- The scheme overran the budget, meaning other planned schemes could not go ahead unless local councils contributed money.
- Properties that are at risk from flooding in places like Wraysbury have high insurance costs.

What are the environmental issues?

- It causes increased flooding at the point where the relief channel joins the main Thames channel.
- Concrete weirs are ugly, especially when exposed out of flood.
- Algae collects behind the weirs, which disrupts the natural ecosystem.

CHECK IT!

1 State the three fluvial processes that occur in a river.

2 Name the four different types of river erosion.

3 Explain how and why material is deposited.

4 Explain how a waterfall is formed.

5 Give three facts about how meanders are formed.

6 Explain how floodplains are formed.

7 Explain how levées are formed.

8 Give two human factors and two physical factors that increase the risk of flooding.

9 a Explain the difference between hard and soft engineering.

  b Give an example of each.

10 a Explain how the Jubilee River scheme has helped to reduce flooding in the Windsor area.

  b Give three reasons why people have objected to the scheme.

Glacial landscapes in the UK

Processes in glacial environments

Figure 26 A glacier

During the last ice age, much of the UK was covered in snow and ice. Ice played a huge part in shaping the physical landscape of the UK. Gravity caused large bodies of ice called **glaciers** to move and flow from north to south.

Glaciers carry out three main processes:

Erosion and weathering: the wearing away of material by ice and water. → **Transportation:** the movement of material by ice. → **Deposition:** the dropping of material by ice and water.

Weathering

The main type of weathering in a glacial environment is **freeze-thaw weathering**.

1. Water seeps into crack in the rocks.

2. The water freezes at night and expands by 9 per cent.

3. The expansion of the ice causes the crack to weaken and crack further.

④ The water melts during the day and seeps down into the wider crack, then freezes again at night causing further cracking.

⑤ This process is repeated over time and large areas of rock can shatter due to this freeze-thaw process.

Erosion

As gravity moves the glacier down the mountainside at a very slow rate, it still allows for two different erosion processes to occur:

- **Plucking:** as the ice moves, the meltwater freezes and bonds to the rocky surface. As the glacier moves, it 'plucks' away pieces of rock leaving behind a jagged surface.

- **Abrasion:** the rocks that are carried in the base of the glacier are dragged along the surface and scour and scratch to leave a smooth surface. When large boulders are dragged along, they can leave large scratches called striations.

Transportation

Glaciers move due to meltwater lubricating the base of the glacier and allowing it to slip downhill, called basal slip. If the movement occurs high up in the valley, it can be a curved movement called rotational slip.

Material transported by a glacier is called moraine. Moraine can be transported by the glacier in three ways:

① In the ice.

② On the surface of the ice.

③ Below the ice.

The front of the glacier is called the snout. The snout pushes material along with sheer force – this is called bulldozing.

Deposition

Deposition occurs when a glacier starts to melt:

- Most deposition occurs where most melting occurs – at the snout.

- As a glacier retreats, it leaves behind lots of broken rock pieces called till or boulder clay.

- Meltwater rivers from a glacier can carry material away and deposit it elsewhere. This material has been worn down by the process of attrition to make much finer, sandy material called outwash.

Glacial landforms

Glaciers have distinct landforms caused by erosional processes.

Corrie

A corrie (also known as a cirque or cwm) is a large hollow found on the mountainsides where glaciers began.

DO IT!

Glacial processes

1 Draw a diagram to show the process of freeze-thaw weathering.

2 Describe two processes of erosion by a glacier.

3 Explain how material is transported by a glacier.

Figure 27 Corrie formation

| Snow gathers in a sheltered hollow on a hillside and compresses to become a dense mass of ice. | → | Freeze-thaw weathering enlarges the hollow and more snow collects and turns to ice. | → | Rotational slip and continued erosion by abrasion makes the hollow bigger and deeper. | → | A raised lip forms at the front of the corrie as less erosion occurs here. A tarn is a lake in a corrie left after the glacier melts. |

Arêtes and pyramidal peaks

- An **arête** is a steep ridge that forms when two corries are back-to-back.
- A **pyramidal peak** occurs when three or more corries have formed on a mountain and weathering and erosion creates a sharp peak.

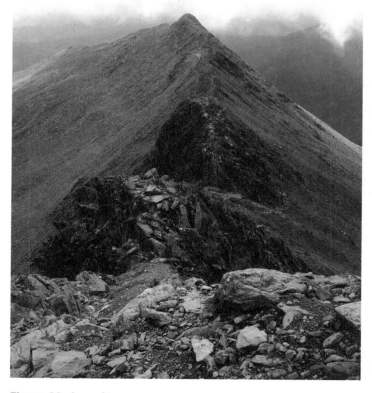

Figure 28 An arête

Glacial valley landforms

- **Glacial trough:** a steep-sided, wide-bottomed U-shaped valley created by erosion by a glacier.

- **Truncated spurs:** left behind when a glacier cuts through the old interlocking spurs of a river.

- **Hanging valleys:** a glacial tributary that was not as deep as the main glacier is left hanging in the new valley wall.

- **Ribbon lakes:** long, thin lakes left behind when a section of the valley was overplucked and made deeper by the glacier.

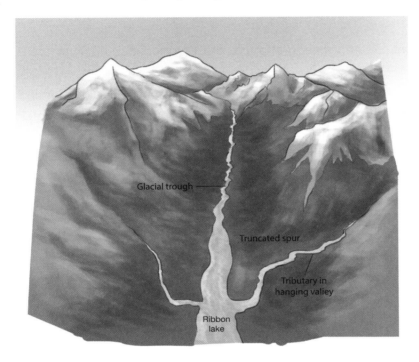

Figure 29 Glacial valley landforms

Glaciers also have distinct landforms caused by transportation and deposition processes.

Moraine

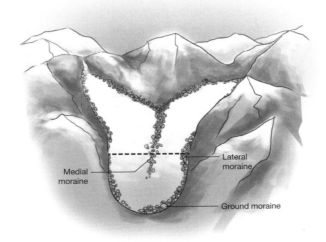

Figure 30 Cross section showing moraine

Corries, arêtes and pyramidal peaks

Write a paragraph to explain how corries, arêtes and pyramidal peaks are formed. Use all of the following words:

- erosion
- weathering
- rotational slip
- moraine
- abrasion
- freeze-thaw.

Moraines are the different rock deposits left behind by a glacier:

- **Lateral moraine:** moraine that forms on the edges of a glacier, fallen from weathering and erosion of the valley sides.

- **Medial moraine:** when two glaciers meet, the two lateral moraines meet and form a ridge of rock debris in the middle of the glacier.

- **Ground moraine:** material that is lodged underneath the glacier and left behind when it melts.

- **Terminal moraine:** an enormous pile of material that builds up at the snout of the glacier.

Drumlins

Drumlins are smooth egg-shaped hills, composed of till. They are formed when:

- a mound of material is deposited by a glacier as ground moraine
- the moving glacier shapes and streamlines the ground moraine into a drumlin shape.

Drumlins have a blunt end facing up the valley and a pointed end facing down the valley. This shows the direction the glacier was travelling.

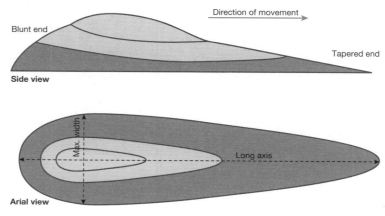

Figure 31 Drumlin

Erratics

An **erratic** is a large boulder that differs from the size and type of rock local to the area in which it is found. It is out of place in the landscape and stands out like a sore thumb. By studying the geology, you can find out where the erratic has come from and, therefore, where the glacier had been.

Economic activities in glaciated areas and management strategies to reduce land-use conflicts

Economic opportunities in glaciated areas

Tourism

- Impressive glacial landscapes attract tourists who enjoy outdoor activities.
- Tourism can provide job opportunities for thousands of people, such as in Aviemore near the Cairngorms, Scotland, which is one of the UK's main mountain activity centres.

STRETCH IT!

Glacial features

Draw your own cross section through a glacier. Annotate it to show the main features of a glacier. You should include the different types of moraine, hanging valleys, truncated spurs, corries, arêtes and pyramidal peaks.

NAIL IT!

Glacial features

Make sure you can describe the different features created by the erosion, transportation and deposition carried out by a glacier. You should be able to explain how they are formed using geographical terminology in your answer.

Farming

- In upland areas, glaciers stripped away the soil and vegetation, leaving thin acidic soils. This land is used for grazing as sheep can endure the cold, wet and windy weather and the poor vegetation.
- In the valleys, the soil is thicker due to deposition. Crops grown in glacial troughs are cereals and potatoes as the flat-bottomed valleys are perfect for farming machinery.
- Lowland glaciated areas are often very fertile due to being covered with a layer of deposited till. This makes them perfect for intensive farming.

Forestry

- Forestry can be found in upland glaciated areas.
- Conifers are adapted to the thin acidic soils and can be planted on the steep slopes.

Quarrying

- In upland glaciated areas, the rock is hard and resistant, which can be quarried and used in the construction industry.
- In the lowland areas, sand and gravel would have been deposited. This is also important in the construction industry, as sand can be made into cement and gravel into concrete.

Conflict in glaciated areas

Economic activities and development in glaciated areas can lead to conflict.

- **Tourism**: this can lead to conflict between local people and tourists as there may be increased traffic congestion.
- **Quarrying**: this is good for economic opportunities, but the process can lead to pollution of the surrounding land and rivers and can spoil the natural landscape.
- **Wind farms**: upland glaciated areas are a good location for wind farms. They can cause conflict with the local people because they may spoil the natural landscape, reduce the number of tourists and affect house prices by having an impact on the view.
- **Reservoirs**: building reservoirs can cause conflict as large amounts of farmland may have to be flooded to build the reservoir.

In glaciated areas, there needs to be a balance between developing the area for economic opportunities and conserving the natural environment. This may be done in various ways:

- **Limiting tourists**: having a maximum number of tourists who are allowed to visit an area at one time so it does not exceed carrying capacity.
- **Signs**: these can instruct tourists on which areas are out of bounds.
- **Seasonal closure**: some areas will be closed during the winter months to allow them to recover.
- **Restricted activities**: camping, horse riding and the use of motorbikes may only be allowed in certain areas.

DO IT!

Economic opportunities in glaciated areas

Create a revision poster to show the economic opportunities in upland and lowland glaciated areas.

NAIL IT!

Economic opportunities in glaciated areas

Economic opportunities are important for people living in glaciated areas. Make sure you know why it is important to have a range of different economic activities.

DO IT!

Create a mind map to show all the conflicts that could arise in glaciated areas and how they can be managed (see the case study on page 86).

STRETCH IT!

Develop an argument as to whether you think it is more important to develop a glaciated area for economic opportunities or conserve the natural environment.

DO IT!

Managing tourism in the Lake District

1 Using the case study here, or the one you have studied as part of your course if it is different, categorise the social, economic and environmental impacts.

2 Create two spider diagrams showing the advantages and disadvantages of tourism in the Lake District.

Case study

Managing tourism: The Lake District

Attractions for tourists

- Water sports, cruises and fishing, e.g. Lake Windermere and Ullswater.
- Walking and mountain biking.
- Adventurous activities, such as rock climbing and abseiling.
- Beatrix Potter and William Wordsworth lived and wrote in the Lake District.
- Idyllic towns and villages to visit.

Impact of tourism

- 40 000 people live in the Lake District. In 2014, 16.4 million tourists visited the area.
- Adventure tourism provides new business opportunities for locals.
- Jobs created by the tourist industry tend to be seasonal and poorly paid.
- Traffic congestion.
- Footpath erosion and littering in popular tourist sites.
- Increasing house prices due to people buying second homes in the area.
- £1146 million was spent in the Lake District in 2014.
- Pollution from cars can destroy the environment.

Managing the impact of tourism

To reduce footpath erosion:

- An organisation called 'Fix the Fells', supported by the National Trust, maintain and repair the mountain paths.
- The Upland Path Landscape Restoration project has created steps and repaired and resurfaced paths.

To manage traffic congestion:

- Park-and-ride bus schemes have been developed to relieve congestion.
- Speed bumps have been introduced in villages to slow traffic.
- Dual carriageways have been built around the Lake District to relieve congestion and improve access.

CHECK IT!

1 Give a correct definition of the term 'basal slip'.

2 Describe the process of freeze-thaw weathering.

3 Explain how material is transported and deposited by a glacier.

4 List five landforms created by glaciers.

5 Describe how a corrie is formed.

6 Explain how a corrie can lead to the formation of an arête or pyramidal peak.

7 State three different types of moraine and where they are found.

8 Explain how striations, drumlins and erratics can all be used as evidence of glacial movement.

Physical landscapes in the UK

Coastal landscapes in the UK

1 Name three types of mass movement.

2 Describe the process of one type of mass movement.

3 a Name two coastal landforms created by deposition.

   b Give a named example of each.

4 Explain how hard engineering protects the coastline.

5 Describe how a destructive wave erodes the coast.

6 Describe the process of longshore drift.

7 Explain how a stack is formed.

River landscapes in the UK

1 Describe how a river changes from source to mouth.

2 Explain the human and physical factors that increase the risk of a flood.

3 Describe two processes of fluvial erosion.

4 Explain how a gorge is formed.

Glacial landscapes in the UK

1 State three different glacial landforms.

2 Explain how a pyramidal peak is formed.

3 Explain how developing a glaciated area can bring conflict.

4 State three different economic activities in glaciated areas.

5 Describe the advantages and disadvantages of an economic activity in glaciated areas.

6 Discuss the potential conflict that could occur due to the development of glaciated areas.

Urban issues and challenges

Global patterns of urban change

THE EXAM!

• This section is tested in Paper 2 Section A.

• You must know all parts of this topic.

DO IT!

Urban growth

Describe the rate of population growth as shown in Figure 1.

Urbanisation

Urbanisation is the growth in the proportion or percentage of the world's population who live in cities. Urbanisation happens because of a natural increase in the population (**birth rate** minus **death rate**) and from migration into the city.

Rates of urbanisation vary across the world. Richer, developed countries have much slower rates of urbanisation compared to poorer, developing countries.

The urban population is not spread out evenly around the world. Just three countries account for 37 per cent of growth – China, India and Nigeria. Asia has the largest proportion of the world's urban population.

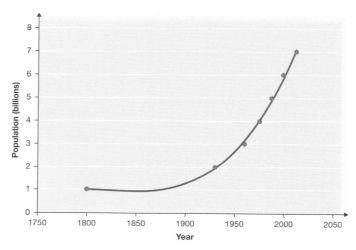

Figure 1 World population growth

Why do cities grow?

There are two main reasons why cities grow:

• **Natural increase**: when the birth rate exceeds the death rate.

• **Rural–urban migration**: when people move from the countryside to the city.

Why do people move from the countryside to the city?

This is due to 'push' factors, which are reasons to leave the countryside, and 'pull' factors, which are reasons to move to the city. It is usually the young and economically active who make the move, aged between 16 and 45. These are the people who are able to work and earn money.

Push factors:

• lack of employment opportunities

• poor harvest leading to **famine**

• **drought** and other climate hazards, such as flooding

- **subsistence farming**, which doesn't bring any extra money to the family
- very few doctors, health care and poor quality, basic education.

Pull factors:

- better paid jobs
- a higher standard of living
- access to better education and health care
- family and friends have already made the move to the city.

Megacities

Megacities are cities with a population of over 10 million people.

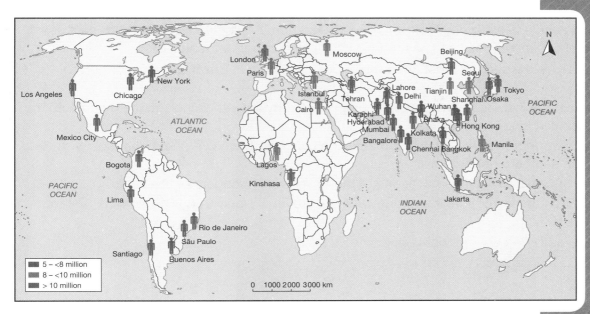

Figure 2 The world's largest cities in 2002

Megacities grow at different rates:

- **Slow growing**: these are found in South-East Asia, Europe and North America, for example, Tokyo and Los Angeles.
- **Growing**: these are found in South America and South-East Asia, for example, Rio de Janeiro and Beijing.
- **Rapid growing**: these are found in South-East Asia and Africa, for example, Mumbai and Lagos.

✓ CHECKIT!

1 What is the difference between urban growth and urbanisation?

2 Describe patterns of global urban change.

3 Give three reasons why people leave the countryside.

4 Give three reasons why people migrate to the city.

5 Describe the location of global megacities.

Urban growth in LICs and NEEs

An emerging city: Rio de Janeiro

Rio de Janeiro is an emerging city in Brazil in South America. It is located on the south-east coast. It is the cultural capital of Brazil and has many exciting and important features:

- Rio de Janeiro is the second most important industrial centre in Brazil and a major port for exports.
- Service industries include banking and finance; secondary industries include the manufacturing of chemicals, clothing and furniture.
- The city is surrounded by mountains and amazing beaches, making it one of the most visited places in the southern hemisphere.
- The Statue of Christ the Redeemer is one of the Seven New Wonders of the World.
- Rio de Janeiro has hosted major sporting events – the FIFA World Cup 2014 and the Summer Olympics 2016.

Rio de Janeiro is divided up into four zones:

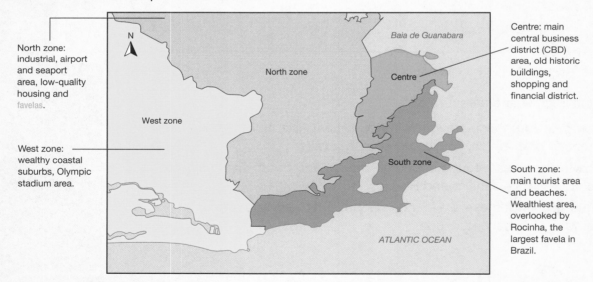

Figure 3 Location map of Rio de Janeiro

DOIT!

Location of Rio de Janeiro

Describe the advantages of this site for the development of Rio de Janeiro or the major city you have studied as an example of urban growth in an LIC or NEE.

Social, economic and environmental challenges

Case study

Social challenges: Rio de Janeiro

Rio de Janeiro has many urban challenges, made worse by the inequalities between the areas of the city. Since 2013, the local government has been working hard to improve and ease the social challenges faced.

Health care

Health care services have been very poor, especially for pregnant women and the elderly in the west zone. Half the population doesn't have a local health clinic.

→ In Santa Maria, medical staff went into the favela and visited people's homes to detect and treat diseases. Infant mortality has fallen and life expectancy has increased.

Education

Only half of the children continued education beyond 14 years due to a shortage of schools, lack of money, shortage of teachers and lack of training and pay for teachers.

→ Authorities have improved access to education by encouraging people to volunteer in schools, giving grants to schools for poor children and providing free sports activities.

Water supply

Around 12 per cent of the population had no access to running water. The water supply system has leaky pipes and water shortages are caused by droughts.

→ New treatment plants and water pipes have been built. Improvements have been made in the favelas so 95 per cent of the population had a mains water supply by 2014.

Energy

Blackouts occur due to power cuts. There is a lack of safety in favelas as people illegally tap into the electricity supply.

→ 60 km of new power lines installed. A new nuclear power generator has been built. Hydro-electric power (HEP) will increase electricity to Rio de Janeiro by 30 per cent.

Figure 4 Life in a favela

SNAPIT! Slum or squatter settlements

Find an image online similar to Figure 4, showing a slum or squatter settlement in the area you have studied. Use it to list the challenges that people who live in the emerging city you have studied might face.

Rank these challenges in a Diamond Nine with the biggest challenges at the top.

NAIL IT!

Social challenges

Make sure you can explain the social problems facing an emerging city such as Rio de Janeiro. You need to be able to describe the solutions to the social challenges in health care, education, water supply and energy.

DO IT!

Economic challenges

For Rio de Janeiro, or the city you have studied:

1 State three economic challenges.

2 State three economic opportunities.

3 Suggest three solutions to the economic challenges.

Case study

Economic challenges and opportunities: Rio de Janeiro

As the city's industrial areas have grown, they have brought economic prosperity to Rio de Janeiro and have helped to improve the infrastructure of the city, with roads, transport and services all being developed. Large companies have been attracted to the city and this has created more economic opportunities and a more **formal economy**.

Positive effects of economic growth in Rio de Janeiro

- It is the second most important industrial city in Brazil.
- The city provides 6 per cent of the whole country's employment.
- It has one of the highest incomes per person in the country.
- Types of employment include: oil, retail, finance, steel, construction, tourism and manufacturing.

Negative effects of economic growth in Rio de Janeiro

- A recession hit in 2015, which increased unemployment.
- There is a wide gap between the rich and poor in terms of wages.
- There is high unemployment in the favelas.
- Most people in the favelas work in the **informal sector** (drivers, street vendors, maids).
- Informal jobs are lower paid, have irregular hours and no contracts or insurance.
- Crime rates are high in the favelas, especially drug trafficking, murder, kidnapping and robbery.

Solutions to the economic problems

- Education is used to reduce youth unemployment by setting up training programmes.
- Free child care is provided for young parents so they can go back to school.
- Police Pacifying Units have been set up to reduce the drug dealing in the favelas.
- Police have taken back control of the worst crime-ridden areas, especially favelas near the Olympic stadium.

NAIL IT!

Economic challenges

You need to be able to describe and explain the economic challenges that urbanisation in Rio de Janeiro or the city you have studied can bring, but also how it brings opportunities. You should also offer solutions to the economic challenges.

Case study

Environmental challenges: Rio de Janeiro

Urbanisation creates a large number of environmental challenges and problems:

- Heavy pollution due to industrialisation.
- Dereliction due to the deindustrialisation of the steel industry.
- Squatter settlements built on hillsides and sprawling out of the edge of the city.
- Urban sprawl along the beautiful coastal areas.
- Smog due to air pollution.

Air pollution and traffic congestion

Rio de Janeiro is the most congested city in South America. Smog due to exhaust fumes from heavy traffic and congestion mixes with mist from the Atlantic Ocean. Air pollution causes thousands of deaths every year.

Causes of air pollution

- A lack of flat land for building.
- Improved transport links.
- Tunnels being built through mountains.
- A 40 per cent increase in car ownership.
- People thinking that car travel is safer.

Solutions

- Expansion of the Metro system.
- Building of toll roads to ease congestion.
- Coastal roads made one-way at busy times to ease traffic flow.

Water pollution

Guanabara Bay is very polluted, which is threatening wildlife and causing a decline in fish populations. It could also damage tourism at Copacabana and Ipanema beaches.

Causes of water pollution

- Polluted rivers flowing into the bay.
- Run-off from open sewers in the favelas.
- 200 tonnes of raw sewage pumped into the bay daily.
- 50 tonnes of industrial waste pumped into the bay every day.
- Oil spills.
- Ships emptying fuel tanks into the bay.

Solutions

- Overseas aid to fund reduction in sewage.
- The building of 12 new sewage works.
- Ships fined for dumping fuel.
- The installation of 5 km of new sewerage pipes.

Waste pollution

Waste in the favelas is dumped as it is too difficult to remove due to the narrow lanes and steep hills. The dumped waste gets into the water supply. This encourages rats and diseases like cholera.

Solution

A power plant that runs on biogas produced from rotting rubbish has been built. It consumes rubbish and produces electricity for the favelas.

DO IT!

Environmental challenges

1 Write three sentences that describe the environmental challenges in Rio de Janeiro or the emerging city you have studied.

2 Explain a solution to each problem.

Challenges and improvements to squatter settlements

Challenges of squatter settlements

Favelas (the Brazilian word for squatter settlements) have grown up in and all around Rio de Janeiro. People have migrated from the north-east of Brazil and the Amazon regions as they are looking for a better way of life.

Favelas are illegal settlements – people have built homes on land that they do not own. They are areas of social deprivation.

> ### Case study
>
> ### A squatter settlement: Rocinha, Rio de Janeiro
>
> Rocinha is the largest favela in Brazil, with a population of approximately over 120 000. It is built on a very high steep hillside overlooking the wealthy areas of Rio de Janeiro.
>
> There are many challenges of squatter settlements like Rocinha.
>
> **Construction**
>
> - Houses are poorly constructed from basic materials such as corrugated iron and plastic sheets.
> - Buildings are built on steep slopes and can be affected by landslides during floods.
> - The steep slopes and tightly packed buildings limit access by road.
>
> **Services**
>
> - Many homes did not have access to running water prior to 2014.
> - Electricity is often obtained through illegal connections.
> - Sewers are open drains that spread disease.
>
> **Unemployment**
>
> - There are high unemployment rates.
> - Most people work in the informal sector, which is poorly paid so incomes are low.
>
> **Crime**
>
> - Crime rates are high in the favelas, especially murder rates and drug-related crime.
> - There are high levels of mistrust of the police.
>
> **Health**
>
> - A high population density makes disease spread easily.
> - Infant mortality rates are high.
> - A high level of waste and open sewers leads to the spread of disease.

DO IT!

Life in a squatter settlement

Link challenges together to extend your points in an exam question. Complete these sentences:

1 In squatter settlements, waste is dumped in open sewers which…

2 A high population density in the squatter settlement causes…

3 Construction on steep hillsides leads to…

STRETCH IT!

Squatter settlements

Squatter settlements are found in megacities all over the developing world. Are there variations in squatter settlements found in Africa, South America or Asia? You should compare the quality of life in squatter settlements in Kibera (Nairobi, Kenya), Dharavi (Mumbai, India) and Rocinha (Rio de Janeiro, Brazil).

Case study

Managing urban growth: the Favela Bairro Project in Rio de Janeiro

The government in Rio de Janeiro has begun an improvement plan to upgrade the quality of life in the favelas. Some of the things they have done include:

- rehousing people in new basic housing
- developing new areas of the city where people can be rehoused
- enforcing evictions in some favelas to allow for redevelopment
- introducing self-help schemes to redevelop the existing favelas
- developing activity programmes for youths to stop them getting into crime.

Figure 5 Self-help housing

NAIL IT!

Squatter settlements

Squatter settlements develop on the edge of megacities in developing countries all over the world. Make sure you can explain why they have formed and what challenges living in a squatter settlement can bring for the people of cities like Rio de Janeiro.

STRETCH IT!

Favelo Bairro Project

Consider whether or not schemes like the Favela Bairro Project have been a success. Make a list of the advantages and disadvantages of the scheme you have chosen.

DO IT!

Challenges and solutions for squatter settlements

Draw a revision poster showing the challenges and solutions to living in a squatter settlement in the emerging city you have studied.

NAIL IT!

Improving squatter settlements

Make sure you can explain how the government in the city you have studied has improved life in the squatter settlements.

A site and service scheme

This US$300 million 'slum to neighbourhood' project integrated existing favelas into the fabric of the city through upgrading the service and infrastructure. The local authority also provided land and services for the people to build their own homes.

- Hillsides are secured to prevent landslides.
- Credit is provided for building materials to be purchased.
- Water pipes and electricity cables supply running water and electricity to houses.
- Health centres, schools and leisure facilities are built nearby.
- A Police Pacifying Unit is set up to reduce crime.
- Roads are paved to allow better access.

SNAP IT! Self-help schemes

Take a picture of Figure 5 on page 95. Use this to help you remember how the local authorities and local people work together to improve life in a squatter settlement.

CHECK IT!

For Rio de Janeiro, or the emerging city you have studied:

1 Describe its location.

2 State three reasons why the city you have studied is an important emerging city.

3 Describe the different zones of the city you have studied.

4 State why squatter settlements build up in emerging cities.

5 Describe the problems of living in a squatter settlement.

6 Explain how local authorities have improved the quality of life in the squatter settlements you have studied.

Urban change in the UK

UK population distribution

In 2015, the UK population was 65 million with more than 83 per cent living in urban areas. The UK population is predicted to continue to rise.

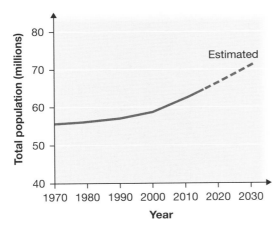

Figure 6 UK population growth

DO IT!

UK population growth

Describe how the population size of the UK has changed over time.

Graphical skills

Describing graphs

When you describe information in a graph you should always:

- Describe the general pattern shown in the graph, for example, the population of the UK has increased.
- Give key figures and data from the graph, for example, from 1 billion in 1800 to 7 billion in 2015.
- Refer to any anomalies or significant patterns, for example, the growth was slow up to 1960, then the population increased rapidly.

The UK population is unevenly distributed, with 83 per cent living in urban areas. These areas developed due to the industrial revolution, at which time people moved to the cities where there was work in factories close to deposits of raw materials.

The biggest cities grew in importance because of their positions as capital cities: England and UK (London), Wales (Cardiff), Scotland (Edinburgh); and Northern Ireland (Belfast). Birmingham grew in importance due to its central location, and some cities such as Liverpool and Bristol developed because of their coastal location for trade.

DOIT!

UK population distribution

Describe the distribution of cities across the UK. How do you think cities may change in future? Which ones will grow and which ones will decline?

Figure 7 Major UK cities

Case study

A major UK city: London

London's importance and location

London is located in the south-east of England on the River Thames. It was first established as a settlement by the Romans in AD 43.

Why did London grow?

- The Thames is a tidal river and allowed for London to become a port.

- London became a centre for trade and commerce.

- It was possible to build bridges across the River Thames, so both sides could be accessed.

- When the port declined, London still stayed a transport hub.

Figure 8 Location of London

- Road and rail networks all focus on London.
- Two major airports at Heathrow and Gatwick are close to London.

London's national and international importance

- London is the capital of England and the UK.
- It is the wealthiest city in the UK and the gap in wealth is widening.

Why is London a world city of global importance?

- The London Stock Exchange is one of the most important in the world.
- London is the home of the headquarters of many major international companies.
- The city is an international centre for media, research, education and culture.
- It is a big attractor for international investment.

NAIL IT!

Location and importance

You need to be able to describe and explain the location and importance of London, or the major city you have studied, in the UK and within the wider world.

Location

For London, or the major UK city you have studied:

1 Describe the advantages of the city's site.

2 Explain why the city was able to become such an important city.

3 Describe the advantages of living in the UK city you have studied.

How urban change has created opportunities

Social and economic opportunities

Impacts of national and international migration on London

Cultural diversity

- London is the most culturally diverse city in the UK.
- Thirty-seven per cent of the population were born outside of the UK.
- Less than half the population of London is White British.
- The largest numbers of migrants have come from India, Jamaica, Nigeria and Eastern Europe.

Recreation and entertainment

- Cultural diversity in London has created opportunities for entertainment and recreation.
- Some areas of London, such as Shoreditch, have developed as a result of the cultural diversity.
- Old Shoreditch has transformed into an eclectic area of museums, art galleries, restaurants, cafes, pubs and Spitalfields market.
- Media, creative and high-tech industries have developed on the sites of old industry.

Employment

- Areas of old heavy industry, like the London Docklands, went into decline from the 1970s onwards.

- The London Docklands area was regenerated and more financial sector industries located here, with office blocks becoming home to major international banks.

- The majority of people in London are now employed in the service sector: in retail; finance; education; the arts; health; and administration.

Transport

- London has an **integrated transport network** – a mix of trains, tubes, buses and cars.

- Driving in London has declined due to increased congestion and the introduction of the Congestion Charge.

- Improvements are being made to the tube and bus networks.

- A new system of cycle lanes and cycle superhighways have been developed across London's main commuter routes.

- A new rail service called Crossrail is being built to link Shenfield in the east with Reading in the west.

- Crossrail aims to reduce journey times across London, increase the number of passengers and improve the journey for those who commute.

STRETCH IT!

London transport

Consider the costs and benefits of the new cycle superhighways and the new Crossrail service for commuters in London.

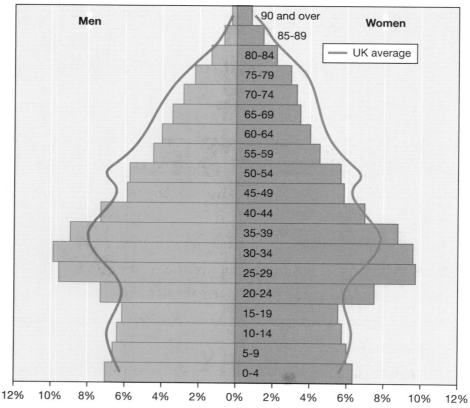

Figure 9 Population pyramid, London, 2011

Environmental opportunities

Urban greening: London

Forty-seven per cent of London is green space! This makes it one of the world's greenest cities. There are lots of parks in London, ranging from the big royal parks, such as Hyde Park, to small neighbourhood parks run by the local councils.

Having lots of green space in cities is good because:

- trees provide oxygen and act as a carbon sink for carbon dioxide
- trees and grass reduce the danger of flooding, as they slow down surface run-off to rivers
- parks and green spaces provide important habitats for wildlife in cities
- parks help to keep people healthy by providing areas for exercise and play.

Urban greening is about how we can maintain, increase and protect our green spaces in cities. In London, it is about protecting the green spaces we already have.

A new plan to increase green space in London is the development of the Garden Bridge. This is a plan to build a new bridge across the River Thames that will be a garden with trees and plants to bring environmental benefits to the city, plus the added economic benefit of being a new tourist attraction.

DO IT!

Social, economic and environmental opportunities

Create a mind map to show the social, economic and environmental opportunities in London or the UK city you have studied.

How urban change has created challenges

Social and economic challenges

Case study

Urban deprivation and inequalities: London

There is a great diversity across London in levels of deprivation. Some areas, such as Kensington, Chelsea and Richmond upon Thames, are very wealthy, have high life expectancy, good exam results and high levels of employment and income. This is a contrast to boroughs like Newham and Tower Hamlets, which are among the poorest in London, with higher unemployment, lower wages, lower life expectancy and poorer exam results.

	Richmond upon Thames	Newham
Life expectancy	83 years	77 years
Unemployment	3.8%	8.5%
% pupils achieving 5A*–C grades at GCSE	70%	57%
Average household income	£46 000	£28 000

Table 1 Differing levels of deprivation in Richmond upon Thames and Newham

Figure 10 London borough of Richmond upon Thames – a non-deprived borough

Figure 11 London borough of Newham – a deprived borough

DOIT!

Social and economic challenges

Study the information in Table 1 and Figures 10 and 11. Compare the two boroughs and describe the differing levels of deprivation. Refer to the figures and data in your answer.

Environmental challenges

Space for building

Many of our major cities have a growing population, which creates a big challenge in terms of housing. New homes need to be built but there is a lack of space upon which to build them. There are two options for building:

- **Greenfield sites**: areas of land that haven't been built on before, such as farmland in the rural–urban fringe. This means that there is urban sprawl, where the city spreads into the countryside. The benefit of this is that land is cheaper as it is further away from the city centre and there is no demolition needed in advance of building. However, valuable farmland will be lost, plus wildlife habitats.

- **Brownfield sites**: areas in the city that have been built on previously and are now derelict. This land might be contaminated from previous industrial use and the old buildings may need demolishing first. There is plenty of brownfield land available from the decline of old industries, but it is expensive due to being in the city. Building on brownfield sites does help to redevelop run-down areas of the city and reduces the need for urban sprawl.

DO IT!

Environmental challenges

Create a table of advantages and disadvantages of building on greenfield and brownfield sites.

Case study

Environmental challenges: London

Green belt

- London is surrounded by an area of green belt land. This is land that has strict planning controls designed to prevent urban sprawl and preserve open spaces around the city.
- There is increasing pressure on the green belt land around London for building houses, as this land is cheaper.
- Demand for housing has meant that towns and cities around London outside the green belt are increasing in size and are now experiencing urban sprawl as new housing developments are built.

Air pollution

- Air pollution in London is created principally by exhaust fumes from vehicles.
- London regularly breaks EU regulations on air quality due to high levels of nitrogen dioxide.
- One way to reduce air pollution in London is by the creation of the new cycle superhighways to encourage more people to cycle rather than drive to work. Cycling has increased from one to 15 per cent over the last 50 years.

Waste disposal

- The majority of London's waste is now recycled, but a quarter is still sent to landfill.
- Landfill is harmful to the environment as it produces methane in the atmosphere, which is a greenhouse gas.
- The aim is for landfill to be reduced to zero by 2030 and to increase the amount of recycling.

Urban regeneration

Why did the area need regeneration?

Urban regeneration: Newham and the Lower Lea Valley

Newham and the Lower Lea Valley in East London was chosen as the site for the 2012 London Olympics.

This site was chosen for a number of reasons:

- There was plenty of derelict land available from previous industrial use and so it was a brownfield site.
- There were high levels of deprivation as Newham is one of the poorest areas in London.
- There were high levels of unemployment.
- It has good transport connections, including the international rail terminal at Stratford.
- There are high levels of cultural diversity.

However, there were issues with the site:

- Areas of land had to be bought up and put into the ownership of the Olympic Development Authority.
- There were some protests over the compulsory purchase of land.
- Old industrial sites had to be cleaned up due to contamination.
- Electricity pylons had to be removed to bury the cables underground.
- The cost of redeveloping the site totalled £9.3 billion.

Figure 12 Olympic regeneration

Was the regeneration a success?

Case study

The Olympic legacy: London

- Queen Elizabeth Olympic Park has open spaces and parkland.
- The Olympic Stadium, Aquatics Centre and Velodrome are all being used by local clubs and people.
- Westfield Stratford City provides 10 000 jobs.
- East Village – the Olympic village now offers over 2800 homes plus shops, entertainment, parkland and a newly built school.
- Here East media centre is a hub for creative industries providing 5000 jobs.
- London Legacy Development Corporation (LLDC) has been set up to continue the development of the area.
- The LLDC has long-term plans in place until 2030.

NAILIT!

Olympic regeneration

You need to be able to explain why Newham and the Lower Lea Valley, or the area you have studied, was in need of regeneration and assess how successful the regeneration scheme was.

DOIT!

Olympic regeneration

Answer the following three questions either with regard to the regeneration of the UK area you have studied or the Olympic regeneration.

1 List three reasons why Newham and the Lower Lea Valley was chosen as the site for the Olympic regeneration project.

2 Create a table of costs and benefits of the Olympic regeneration project for Newham and the Lower Lea Valley.

3 Classify the changes into economic, social and environmental categories.

CHECKIT!

1 State the percentage of people who live in urban areas in the UK.

2 a Name the largest cities in the UK.

   b Give a reason why location has helped the development of cities, such as London in the UK.

3 Explain why London, or the city you have studied, is a city of global importance.

For London, or the UK city you have studied:

4 State three ways that transport for the city has been improved.

5 Describe the cultural mix of the city's population.

6 Explain how urban greening has improved environmental quality in the city you have studied.

Either with regard to the urban regeneration you have studied or the Olympic regeneration:

7 Give a correct definition of the term 'regeneration'.

Urban sustainability in the UK

Urban sustainability

A city today needs to be sustainable so it can meet its needs but still be able to provide for, and meet, the needs of future generations. Cities have a series of inputs and outputs that must be managed. The ecological footprint of a city is the area of land that is required to produce all the inputs needed and to dispose of the outputs.

Inputs	Outputs
Food imported from the countryside or from abroad.	**Waste** needs to be sent to landfill sites outside the city.
Water provided from rivers and stored in reservoirs.	**Sewage** needs to be treated before it can re-enter the cycle.
Energy from burning fossil fuels.	**Pollution** in the air spreads beyond the city.
Resources like building materials and anything else we consume.	

Table 2 Inputs to and outputs from a city

DO IT!

East Village sustainability

Create a mind map to show how East Village, or the area you have studied, is a sustainable community.

Case study

A sustainable community: East Village

East Village was created as part of the Olympic legacy in East London. It is part of the sustainable plan for the Olympic development to have a long-term function after the games ended.

How is it sustainable?

- Affordable housing – lower rents – affordable for local people.
- Green spaces – 10 ha of parkland with trees and ponds.
- Energy-efficient housing – high levels of insulation in high-rise apartment blocks.
- Green roofs – to encourage wildlife and slow water run-off.
- Transport – bus, trains and tube service, plus walking and cycle routes.
- Public services – school and health centre for local people.
- Local shops – small independent businesses to keep money in the local economy.
- Water – usage is 50 per cent less than in most urban areas and rainwater is filtered and recycled for flushing toilets.
- Energy – usage is 30 per cent less than that of an average urban area. Combined heat and power systems (CHP) use biomass to produce heat and hot water.

Sustainable urban transport

 Case study

A sustainable transport network: Bristol

Bristol has developed a new sustainable transport strategy using a variety of different strategies:

- Cycling – the Bristol Cycling Strategy provides a network of cycle routes that aims to keep cyclists out of heavy traffic.

- Public transport – MetroWest is a metro rail service linking the surrounding towns with Bristol City Centre. The MetroBus provides rapid transit buses to shorten journey times across the city.

- Other sustainable strategies – 30 km/h speed limits in built-up areas are enforced to make it safer for cyclists and pedestrians. Several residents-only parking zones have been designated to reduce traffic congestion. A park-and-ride scheme has been set up to allow people to leave their cars on the edge of the city and travel in by bus to reduce congestion.

Sustainable transport

Give three ways that planners in Bristol, or the area you have studied, have made transport more sustainable and explain how these work to reduce traffic.

 # CHECKIT!

1 Define the terms 'sustainable' and 'ecological footprint'.

2 Give three ways that East Village is an example of sustainable urban living.

3 Explain how Bristol has developed its transport network to be more sustainable.

Global patterns of urban change

1 Give a correct definition of the term 'megacity'.

2 Describe how the global population has changed over time.

3 Explain why people move from the countryside to urban areas in developing countries.

Squatter settlements

1 Describe the social challenges of living in a squatter settlement.

2 Describe the economic challenges of living in a squatter settlement.

3 Explain how the quality of life in squatter settlements can be improved.

Urban change in the UK

1 Give a correct definition of the term 'urban greening'.

2 Give three advantages to building on brownfield sites.

3 Describe the opportunities for recreation and entertainment in London.

4 Describe the difference between two London boroughs or another urban area you have studied.

5 Explain the importance of an integrated transport network in London.

6 Explain the reasons for choosing the Lower Lea Valley site as an area for the Olympic development.

7 Assess the success of the Olympic regeneration scheme.

How urban change has created challenges

1 Give a correct definition of the term 'deprivation'.

2 Give three social and economic challenges in a city like London.

3 Give three environmental challenges in a city like London.

Urban sustainability in the UK

1 Give a correct definition of the term 'sustainable'.

2 State two inputs and outputs of a city.

3 Evaluate the effectiveness of an urban transport scheme you have studied.

4 Using an example you have studied, explain how settlements can be sustainable.

The changing economic world

Economic development and quality of life

Classifying economic development and quality of life

Economic development is the making of money through jobs and businesses and so is measured and classified by amounts of money. This is commonly measured by **gross domestic product (GDP)** per person and also by **gross national income (GNI)**.

It would be expected that the wealthier the people and the more successful the businesses, the more a country would develop and the better the **quality of life** of its people. Quality of life can be judged by social factors such as education, health care, happiness, freedom and gender equality, as well as how much money people have available for spending. However, economic development does not always lead to a better quality of life for everyone in a country straight away.

THE EXAM!

- This section is tested in Paper 2 Section B.
- You must know all parts of this topic.

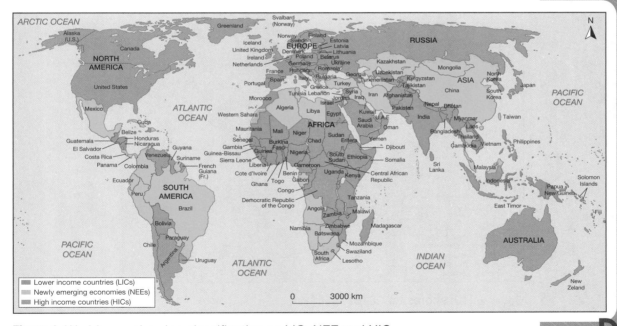

Figure 1 World map showing classification as LIC, NEE and HIC

Economic and social measures

Economic measures

- Gross national income (GNI) per person: takes all of the money made within a country together with all of the money made from investments overseas, minus any debts, and divides it by the total population to find an average of how much each person makes in a year.

DO IT!

Economic development of countries

Copy and annotate Figure 1 to summarise the characteristics of the three economic categories of country:

- low income country (LIC)
- newly emerging economy (NEE)
- high income country (HIC).

NAILIT!

Quality of life

There are several ways of looking at 'quality of life', but perhaps the most important is having access to the basic things that are required in order to stay alive – such as clean fresh water, enough food and a balanced diet, a strong shelter from the weather and for safety. In the modern world, necessary additions may be to have a job to make money and services to help you stay healthy, such as a sanitation system and health care, and education so that you can improve your life. Very few people in LICs are likely to think about fashion clothing and the latest electronic gadgets!

NAILIT!

Measures of development

Make sure that you can define each measure of development. The HDI is considered by most people to be the best composite measure available as it covers social and economic factors.

Social measures

- Access to **safe water**: measures the percentage of the population that is able to get and use clean water. This reflects the amount of money spent on better sanitation systems to prevent sewage mixing with drinking water, or digging and concrete lining more wells.
- **Adult literacy rates**: the percentage of adults in a population who can read and write. This reflects the amount of education received.
- **Birth** and **death rates**: measure the number of babies being born and people dying per 1000 of a population. These reflect the level of education of mothers, investment in pre- and post-natal care, the living conditions and the ability of a health care system to tackle disease.
- **Infant mortality**: measures the number of deaths of babies under one year of age per 1000 live births. This reflects the living conditions and the level of specialist health care available in the country or place.
- **Life expectancy**: measures how many years a new-born child can be expected to live for in a country. This reflects the access to health care and the living conditions in the country.
- **People per doctor**: measures the average number of people a doctor has to look after in a country or place. This reflects the level of education and training of medical specialists and the number of people that a country can train.

Economic and social measures

- **Human Development Index (HDI)**: measures both economic and social factors to try to give an overall measure of development (see Figure 2). It includes life expectancy, literacy levels and GDP.

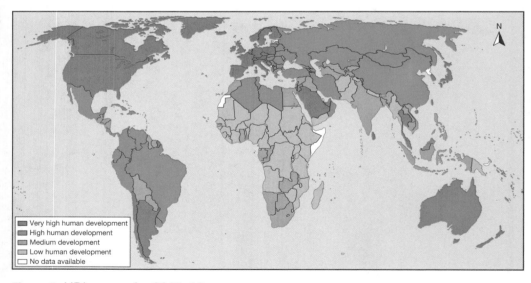

Figure 2 HDI scores for 2015–16

Limitations of economic and social measures

Each measure of development has strengths and weaknesses:

- Economic: GNI per person (or per capita) is a strong measure of wealth within a country and also of the role of the country in international business. However, by creating an average it hides the difference between the very rich and the very poor, and it does not consider the fact that the 'cost of living' varies between countries or that currency values change.

- Social: in low income countries (LICs) and newly emerging economies (NEEs), data collection by governments is difficult due to lack of money, the remoteness of many rural areas and the overcrowding of some urban areas. Therefore, the accuracy of data is variable.

- Economic and social: HDI has been used by the United Nations (UN), a reputable organisation, since 1990, so is a reliable way of measuring a combination of social and economic development. However, like many other measures, it does not consider measures of the natural environment, which is also important to long-term sustainable development. It also omits measures of human rights (e.g. gender equality).

STRETCHIT!

Measuring economic development

Sometimes the Gini index is used to compare the distribution of income amongst a population between countries; this index looks at how wealth is divided amongst a population. Purchasing power parity (PPP) can be calculated to compare how much US$1 would buy in different countries to consider 'cost of living' differences. (The BigMac index works in a similar way.)

 Graphical skills

GNI compared to infant mortality (scattergraph and line of best fit)

Using the data in Table 1, complete the following steps.

1. Draw two axes on graph paper, one for GNI (from 0 to 60) and one for infant mortality (from 0 to 90).

2. Plot the position of each country with a cross on the graph to create a scatter of points.

3. To show the relationship between GNI and infant mortality, draw a line of best fit. To do this, judge where the middle is between the points as you move from left to right across the graph, so that there is an equal number of points either side of your line when you have drawn it.

4. What pattern or trend do you notice from your completed graph?

DOIT!

Measuring development

Plan an answer to a question that asks you to suggest the most reliable way of measuring development.

Country		GNI (US$'000s per capita)	Infant mortality rate (per 1000 live births)
HIC	Chile	21.74	7
	Japan	38.87	2
	Saudi Arabia	54.73	13
	UK	40.55	4
	USA	56.43	6
NEE	Brazil	15.02	15
	China	14.16	9
	India	6.02	38
	Mexico	17.15	11
	South Africa	12.83	34
LIC	Afghanistan	1.99	66
	Chad	2.11	85
	Ethiopia	1.62	41
	Haiti	1.76	52
	Nepal	2.50	29

Table 1 GNI and infant mortality data for selected countries in 2015

Links between population change and development change

As a country develops economically, and people and governments have more money, the culture of the country begins to change. For example, in the past, as countries developed they became more democratic – people had greater freedom such as free elections and freedom of speech. There was also the ability to spend more on education, health care and welfare systems to look after people, which changed birth and death rates (Figure 3).

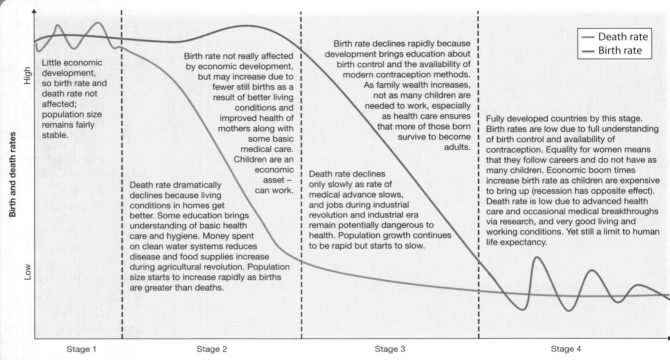

Figure 3 Links between level of development and demographic transition

SNAPIT!

Demographic transition model and economic development

Snap an image of Figure 3, showing the links between population size and birth rates and death rates as a country develops economically over time. There are quite a few linked ideas here.

To help consolidate your understanding, study the image a few times during the revision process and test yourself on what happens in each stage of the **demographic transition model (DTM)**.

Countries that are overpopulated have difficulty providing **resources**, services and jobs for everyone and so their development is slower, for example, Chad. But underpopulation can also be a problem, as there are not enough people to do the work that needs doing, for example, in Japan with its ageing population.

Causes of uneven development

Uneven development has happened on a world scale and also within countries. There is some disagreement about the strength of the causes of uneven development, and there may be a difficult transition time within the development process.

Physical geography

1   **Climate** has had a long-term influence on the development of countries. Low temperatures (temperate climate) allow a wider variety of

crops to grow and livestock farming to take place, so there is a plentiful and reliable food supply; there are fewer fatal diseases and working is much easier as it does not get too hot for the human body. In contrast, extremely hot or cold places are more difficult areas to develop.

2 **Natural resources** are found in certain places, such as fertile soils and coal in temperate areas, or oil in hot areas such as the Middle East and in cold areas such as the Arctic region. Those areas with more natural resources have developed faster than those without.

3 **Location** provides access to other areas for trade around the world. The sea has, for a long time, been the best way of transporting heavy and bulky goods. So those countries or regions with access to the sea have developed faster than those without it.

4 **Natural hazards** may occasionally slow rates of development, and it is not yet known how **climate change** may increase the frequency of weather-related hazards such as floods, **droughts** and tropical storms.

Economic geography

1 **Trading** helps countries to develop as they are able to get the resources they need and sell the products they make. However, not all countries have resources; food exports depend on the weather to produce high yields, or the ability to make products and so some countries remain poorer than those that are able to trade. Not all trade is 'fair', with HICs often gaining and LICs losing in trade agreements.

2 **Foreign direct investment (FDI)** helps countries because businesses from other countries spend money building factories and bringing machinery and new ideas.

3 **Spending on education and health** improves the skills and abilities of workforces, helping businesses to be more efficient and make more money.

4 **Government policies** based on a 'free market' where businesses are able to compete and the best survive have tended to bring more sustained development than in those countries that have heavily supported old inefficient industries. Corruption in some LICs and NEEs has prevented benefits from reaching poor people.

Historical geography

1 **Colonial expansion** by European countries led to many of the poorer parts of the world being politically controlled and economically dominated by the European countries from the 18th century to around 1945. Some say that the rich countries of Europe became richer and the poor countries became poorer because of an unequal relationship in terms of trade and exploitation of resources. There was a lack of investment in colonies and after independence these countries had trouble developing, often made worse by internal conflicts as borders did not match ethnic areas.

2 **Aid** was given by government organisations and **non-governmental organisations (NGOs)** or charities to many poorer countries from 1945 to help them develop. This aid was in several forms, including foreign direct investment, medical help, education help, farming and clean water aid, trading agreements and loans for expensive schemes. Some believe that this made the poorer countries dependent on help instead of allowing them to develop themselves.

DO IT!

Uneven development

Ten causes of uneven development in the world are listed here. Discuss these with a friend. Decide together on a rank order which starts with the cause that you think was the most important in creating the worldwide gap between rich and poor countries that we find today, finishing with the cause that you think was least important.

Make a note of your order as a revision list.

Consequences of uneven development

Uneven development may mean that some people have what they need for life and others do not. Some aspects of development may happen faster than others, such as economic development happening before social development.

SNAPIT!

Cycle of poverty

Snap an image of Figure 4. Use this image to remind you of how many people remain poor in the modern world. Think about the ways in which this cycle could be broken so that people could have a better quality of life.

Disparities in wealth

- Development has not spread wealth evenly to everyone. About 13 per cent of the world's population live on less than US$1.90 a day, but in sub-Saharan Africa the figure rises to about 43 per cent. In 2012, India (NEE) had 22 per cent of the population with less than US$1.90 a day and in South Africa (NEE) the figure was 16.6 per cent. Even in China (NEE) the figure was 11 per cent.

- Many people in LICs and NEEs are stuck in a **cycle of poverty** (see Figure 4), which is difficult to break out of without an input of money. In Haiti (LIC), 54 per cent of the population live on less than US$1.90 a day (2012) and in 2010 the figure for Madagascar (LIC) was 82 per cent.

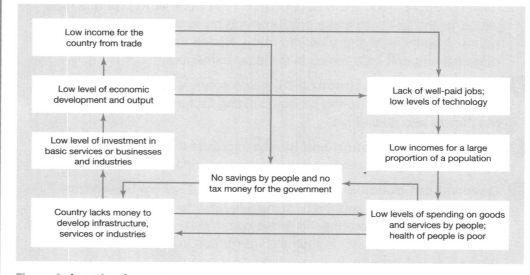

Figure 4 A cycle of poverty

STRETCHIT!

Classification of countries

The World Bank uses the categories 'high income' and 'low income' for countries in its World Development Indicators database, and divides the newly emerging economies (NEEs) into 'lower middle income' and 'upper middle income'. Countries are not fixed in these categories and may move between them as development paths are not steady. For example, India is placed in the lower middle income category but has an income per person about half that of the other NEEs, and more than 20 per cent of the population live on less than US$1.90 a day.

Disparities in health

- In 2015, about 9 per cent of the world's population did not have access to safe clean water and about 32 per cent did not have basic sanitation. These people are more likely to get life threatening diseases. In 2015 in the LICs, 66 per cent of people had access to improved water but only 28 per cent had access to improved sanitation. HICs do not face these problems. Many LICs and NEEs neither have the ability to provide health care in remoter areas nor can afford modern medicines to cure medical problems, as shown by high infant mortality rates (Table 1, page 111) and other health data (Table 2, page 115).

Country		Doctors (per 1000 people)	Incidence of tuberculosis (per 100 000 people)
HIC	Chile	1.0	16
	Japan	2.3	18
	Saudi Arabia	2.5	12
	UK	2.8	12
	USA	2.5	3
NEE	Brazil	1.9	44
	China	1.9	68
	India	0.7	167
	Mexico	2.1	21
	South Africa	0.8	834
LIC	Afghanistan	0.3	189
	Chad	na*	159
	Ethiopia	0.0	207
	Haiti	na*	200
	Nepal	na*	158

*data not available (likely to be very low)

Table 2 Health data for selected countries, 2014

Statistical skills

Dispersion and central tendency

Use the data in Table 2 to carry out the following calculations:

1 Find the **median** for the number of doctors per 1000 people.

2 Find the **mean** for the cases of tuberculosis per 100 000 people.

3 Give the **range** for both sets of data.

4 Find the **modes** for the number of doctors per 1000 people.

5 Place the data for number of doctors into **modal classes** 0 to 0.5, 0.6 to 1.0, 1.1 to 1.5, 1.6 to 2.0, 2.1 to 2.5 and 2.6 to 3.0 (ignoring the na data). Which is the most common **modal class**?

6 Plot the data for cases of tuberculosis on a single-axis scattergraph. Use this to calculate the upper and lower **quartiles** and **inter-quartile range**.

7 Use the data and all the calculations to make conclusions about the quality of life in the three different types of country.

International migration

- **Globalisation** processes such as fast transport and instant communication, as well as the spread of 'Western' culture, have increased awareness of the possibility of a better quality of life. This has made many people in poorer countries want a new and better life, which has encouraged both skilled and unskilled people to move (see Figure 5 on page 116).

• Many people around the world are wealthier than in previous generations, and transport methods are more easily available, so the pull to richer countries and the push from poor countries has increased migration. Refugee numbers in particular have grown, as have illegal migrations in places.

DOIT!

Push and pull factors in international migration

Draw a diagram to show how the push factors in a poor country and the pull factors in a rich country combine to produce international migration.

Figure 5 Flow line map to show recent international migration

STRETCHIT!

'Brain drain' and remittances

LICs/NEEs can benefit or lose because of emigration to HICs. If the most educated and skilled people leave (brain drain) for higher paid jobs in HICs, the poorer countries lose the very people who could help them most to develop the country. However, all types of migrants earn more money in the HICs and usually send money (remittances) back to their families, so helping to break the cycle of poverty and stimulating development by increased local spending.

CHECKIT!

1 Give a definition of the term 'quality of life'.

2 a Outline one economic measure of development.

   b Outline one social measure of development.

3 Using any measures of development, give two differences between LICs and NEEs.

4 a Give the characteristics of the birth rate and death rate at Stage 2 of the demographic transition model (DTM).

   b Explain the state of economic development of a country at Stage 2 of the DTM.

5 Compare the development problems created by overpopulation and underpopulation.

6 a Describe two physical geography causes of uneven development around the world.

   b Explain two human geography causes of uneven development around the world.

Global development gap

Strategies to reduce the development gap

Strategies are often in the form of projects or schemes, which can be large and government-led (top-down) or small and community-led (bottom-up). All aim to promote development through a multiplier effect (see Figure 6).

Investment

This strategy involves putting money into a project.

- The money could be a loan from the World Bank to an LIC or NEE to build **infrastructure** that would help to increase exports, such as better port facilities.

- The money could be from foreign direct investment (FDI) where foreign **transnational corporations (TNCs)** build factories and develop links with local businesses.

- Money could come from international migrants working abroad who send money (remittances) back to their families.

Industrial development and tourism

- One strategy involves an LIC or NEE finding something that it has that is unique, which people in other countries would want to buy. Advantages may be climate, physical resources or natural landscape. This could lead to the development of industries based on the resource or the development of tourism based on a hot and sunny climate, scenery, tropical beaches or wildlife (see case study on Jamaica on page 120).

- However, tourist trends change and resources may be used up, creating uncertainties on the path of development.

NAILIT!

Infrastructure

Make sure that you understand that 'infrastructure' refers to all of the things that need to be in place before a business or industry can operate. At the site of a business or factory, energy and water supplies, a waste disposal and sewerage system and roads will be needed. Away from the site, long-distance roads or railways and ports and airports may be needed. So money is often invested in these things.

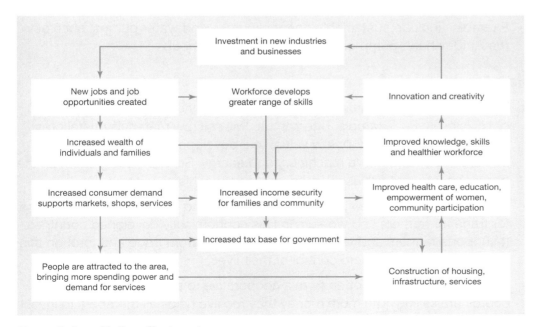

Figure 6 A multiplier effect cycle

DOIT!

Development strategies

Working on your own, or with one or two friends, think about how each development strategy would reduce the development gap. For example, explain how giving aid to an LIC would help it to develop. Write down your ideas.

Development theories

There are theories linked to the development of countries. The Rostow model predicts that countries go through stages, stepping up each time until eventually they become fully developed. In contrast, dependency theory suggests that the rich core countries use the poor peripheral countries to stay rich, maintaining inequality in the world.

Aid

This strategy involves an LIC or NEE getting help from other countries, either directly from the government of another country or from NGOs (international aid).

- The UN has set a target of developed countries donating 0.7 per cent of their annual wealth to help LICs or NEEs. Not many countries reach this target. The aid is usually tied to specific major projects or schemes (top-down approach) with long-term benefits, but some criticise this for not helping the poorest people enough. There may also be corruption in the LIC or NEE, which stops a lot of the money being spent on the project or scheme.

- NGOs such as Oxfam, WaterAid or Practical Action (i.e. charities) collect donations from people in developed countries and use the money in smaller projects and schemes directly linked to poorer people, such as health. (This is described as the bottom-up approach.) Some see this as a more sustainable approach, but a criticism is that it does not help many people.

Intermediate technology

- The most advanced technologies (high-tech) are often very expensive and so not suitable for LICs or NEEs. Many think that improvements to the basic technology found in poorer countries should only be through intermediate technology. This is more appropriate because it is affordable, does not use expensive energy sources, is easy to understand and is easy to maintain and repair.

- This level of technology can be applied to villages and communities or to industry and businesses to make small but significant improvements. Examples include sustainable energy sources and water pumps, such as the Afridev hand pump or the Playpump roundabout.

Fair trade

- In a normal trading situation, a developed country usually gains and an LIC or NEE loses. For example, often an LIC will sell raw materials (metal ores, crops), which are relatively low value, to a developed country but will buy back relatively expensive machinery and energy. So in a trading situation the poorer country actually loses money. Many see this as unfair.

- Fair trade is about obtaining better prices, working conditions and fair terms for trade for farmers and workers in less economically developed countries. It supports farmers and workers in gaining more from trade and through this they are able to take more control of their lives.

- Small-scale producers often form a cooperative to share costs and coordinate selling. The more money they receive helps all members to invest in their farms and businesses and improve the quality of life for their families.

- The UK is the largest buyer of fair trade products (Figure 7).

Figure 7 FAIRTRADE Mark

- The Fairtrade Foundation is an independent non-profit organisation that licenses use of the FAIRTRADE Mark on products in the UK in accordance with internationally agreed Fairtrade standards. When you buy products with the FAIRTRADE Mark, it means that the ingredients in the product have been produced by small-scale farmer organisations or plantations that meet Fairtrade Foundation social, economic and environmental standards. These standards include protection of workers' rights and the environment, payment of the Fairtrade Minimum Price and an additional Fairtrade Premium to invest in business of community projects.

- More than 1.65 million farmers and workers in 74 countries benefit from Fairtrade certification for their products.

Debt relief

- Many LICs or NEEs borrowed money from the World Bank and other sources from the 1960s onwards to help them build projects designed to assist them to develop. However, the economic **recession** of the 1980s slowed development to a point where they could not maintain repayments and interest payments. By the 1990s they had very large debts, and many of these debts have continued into the 21st century.

- 'Debt-for-nature swaps' have been used to reduce the debt of LICs or NEEs. This is where the poorer country agrees to protect part of its natural environment in return for some of the debt being 'wiped out'. One example is a US$21 million swap between the USA and Brazil to protect an area of the rainforest.

- The World Bank and the International Monetary Fund (IMF) introduced the Heavily Indebted Poor Countries Initiative (HIPC) in 1996, which defers repayments until a later date when the poor countries have developed and can afford to repay.

Fairtrade

Next time you visit a supermarket, look out for Fairtrade products for sale with the FAIRTRADE Mark (see Figure 7) and study their labels. Make a note of each product and the country it came from.

Loans and debt

Create a podcast to bring the problems of loaning money to countries to the attention of the World Bank and IMF.

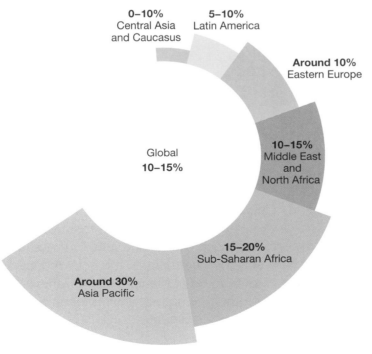

Figure 8 Development and growth trends across the world's microfinance markets in 2016

STRETCH IT!

Obstacles to development

Jamaica faces obstacles to development that are not widely publicised, as they may reduce tourist numbers. For example, there can be **hurricanes**, which cause considerable damage with storm surges and strong winds. Crime rates are also high, and it has one of the highest murder rates in the world, linked to drugs and gang violence.

Microfinance loans

- **Microfinance loans** are small amounts of money lent to poor people by local banks to help break the cycle of poverty. The money must still be repaid with interest, but dependency on aid is avoided.

- Microfinance loans could be used for home improvements, business improvements or education. Loans could be taken out by a group of people or individuals. However, the interest payments could still be too high for poor people and some require a guarantor (someone to pay in case the person receiving the loan cannot), which is not easy to find in communities where everyone is poor.

- The growth of mobile phone use even in sub-Saharan Africa (through the donation of second-hand ones from developed countries) has increased and allowed small businesses, including farmers and fishermen, to obtain micro-insurance and also microfinance loans to make improvements to their businesses.

Case study

Tourism in an LIC or NEE – reducing the development gap: Jamaica

Jamaica is an island in the West Indies, in part of the Caribbean. This world region has an image of tropical islands with blue seas and skies and palm-fringed sandy beaches. Cruise ships move between the islands as well as international airports bringing long-stay visitors. These physical characteristics provide an advantage to develop tourism.

Tourist economy

Tourism has grown from 1.32 million tourists in 2000 and 0.91 million cruise-ship passenger visits to 2.12 million and 1.57 million respectively in 2015. According to the World Travel and Tourism Council, the total direct and indirect income for Jamaica from travel and tourism amounted to 27.2 per cent of GDP in 2014 and they predict that this will increase to 37.5 per cent by 2025. In 2014, there were 82 500 jobs directly linked to tourism (7.3 per cent of total employment) and this is predicted to increase to 130 000 jobs (10.6 per cent of employment) by 2025. However, Jamaica has experienced fluctuations and there was a peak in 2006 before the world recession reduced the number of tourists, although the numbers now seem to have recovered.

Multiplier effect

With jobs, people have more money to spend in local shops and businesses, which allows them to employ people and pay them wages. The government has invested in infrastructure to support tourism through better roads near the tourist resorts. Those with tourist jobs or those who own tourist facilities live in good quality housing. New sewage treatment plants have reduced pollution, and unsightly areas have been landscaped (e.g. Montego Bay). Marine Nature Parks (e.g. Negril) have been established, which attract tourists and protect the natural environment. **Ecotourism** has brought benefits to more remote small-scale tourist businesses.

Development problems

Tourists do not always spend much money outside their 'package deals'. Infrastructure improvements have not spread to the whole island. Some parts of the island do not benefit from tourism at all, with people living in poor quality housing and lacking access to basic services. Too many tourists concentrated in a small area cause footpath **erosion** and pollution.

Year	Tourists stopping (millions)	Cruise passengers visiting (millions)
2000	1.32	0.91
2005	1.48	1.14
2010	1.92	0.91
2015	2.12	1.57

Table 3 Tourist data for Jamaica, 2000–15

Statistical skills

Calculating percentage change

Study Table 3.

1. Calculate the percentage change in the number of tourists stopping in Jamaica between 2000 and 2015.

2. Calculate the percentage change in the number of cruise passengers calling in on Jamaica between 2000 and 2015.

3. What weaknesses are there, or could there be, with the statistics shown in Table 3?

DO IT!

Tourism in Jamaica

Create a revision card that has six bullet points showing how the growth of tourism has helped to reduce the development gap in Jamaica, or the LIC or NEE you have studied, if different.

Hint: do not be too general – include information and facts specific to Jamaica or the area you have studied.

CHECKIT!

1 Describe the multiplier effect.

2 a Explain how giving aid to a country can help it to develop.

 b Compare the suitability of fairtrade and debt relief for helping an LIC to develop.

 c Compare the suitability of microfinance loans and intermediate technology for helping poor people in an LIC to improve their quality of life.

3 For a named LIC/NEE that you have studied:

 a Explain the natural advantages that the country has for developing tourism.

 b Describe the positive contributions that tourism makes towards the development of the country.

 c Discuss whether the development of tourism has been a success for (i) the growth of the country's economy, and (ii) improving the quality of life for the people.

Rapid economic development and change

Rapid economic development and change in an NEE

 Case study

Change in Nigeria

Location and importance

Nigeria is located in West Africa. It has a coastline on the Atlantic Ocean and has borders with four countries: Cameroon to the east, Benin to the west, Niger to the north and Chad to the north-east. It is located just north of the equator and so has tropical rainforest conditions in the south near the ocean but it is much drier in the north, becoming semi-arid towards the border with Niger.

Nigeria is an NEE and one of the economic leaders in Africa. It is expected to become a world top 20 economic nation soon; this is largely linked to oil reserves – it supplies 2.7 per cent of the world's oil, making it the 12th largest producer. It has the largest population in Africa, with 184 million people. This provides a large young adult workforce, and many Nigerians live in different places around the world so the country receives high remittances from those abroad (US$20.8 billion in 2015). Politically, Nigeria contributes armed peacekeeping forces to the African Union and the UN.

Political, social, cultural and environmental context

Nigeria was a UK colony but gained independence in 1960. Like most other African countries, it suffered from a series of civil conflicts with fighting between different factions. The situation has been relatively stable since 1999 and the last two national elections were regarded as democratic. There is still unrest with insurgents in the north (Boko Haram) and in the River Niger delta area where indigenous people (Ogoni) have suffered due to the exploitation of oil. Widespread corruption is holding back the development of the country, much of it linked to the oil industry.

Nigeria is a multi-racial society with many ethnic groups – the main groups are: Hausa and Fulani, Yoruba and Igbo. With the different ethnic groups and a history of external influences, there are many religions, including Christianity (in the south) and Islam (in the north). Over the years, conflicts have often centred on the ethnic divides, and recently a north-south divide has grown due to wealth in the south not spreading to the north – which has allowed Boko Haram (a militant Islamist group) to cause serious trouble.

There is still considerable poverty in the country, which is reflected in the quality-of-life data. The infant mortality rate is 18 times higher than in the UK, the literacy rate is only 61 per cent and the GNI per capita is only 15 per cent that of the UK. It is estimated that 70 per cent of the population live in poverty.

Nigerian music, cinema (Nollywood), literature and sport (notably football) are all well known in Africa and in the world, perhaps through the internet (see Figure 9).

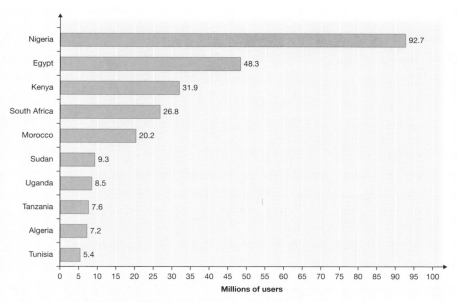

Figure 9 Africa's top ten internet countries in 2015

The environmental conditions in the country have helped to shape its culture, way of life and political context. In the semi-arid north, there are difficult living conditions, but where the savannah grasslands are found, cattle are grazed and a few crops are grown, such as millet. In the centre of the country, there are areas of farmland where it is cooler and wetter, with a denser population around places such as Jos and the capital Abuja. In the south of the country, there is still a lot of forest but also agroforestry with plantations of export crops such as rubber.

Changing industrial structure

Traditionally, Nigeria traded in agricultural products and these still remain important, but with the discovery of oil there was a big push to develop this industry (oil now makes up 91 per cent of exports). However, it is estimated that there are only about 50 years of oil left, so Nigeria will have to find alternatives to keep its economic development going. There is a more balanced economic structure now, but most people still work in agriculture with 25 per cent in the oil industry. Exports are mostly raw materials, which have a lower value than manufactured products (see Table 4 on page 125).

Figure 10 Recent changes to Nigeria's crude oil production

Manufacturing in Nigeria is slowly increasing, with products such as soap, textiles and processed foods. Jobs have been created and people are able to earn better wages. As many people speak English, there may be opportunities to link into global finance and information technology using modern communication methods.

A recent decline in oil prices has affected Nigeria's GNI, which has also prompted increased regulation of remittances to keep exchange rates stable and reduce currency speculation.

The role of TNCs

Due to the relatively settled political situation, there was investment in Nigeria from China, South Africa and the USA. However, the instability in the Niger delta and in the north has restricted the amount of external investment. In 2016, FDI totalled US$887.32 million. Chinese companies are building a 1400 km railway along Nigeria's coast, and developing a new oilfield.

TNCs have located in Nigeria because the workforce there is cheaper and they can be closer to the raw materials needed for their factories. For example, Unilever (producing food, soaps and detergents) is one of about 40 TNCs found in Nigeria. Unilever makes products for export and the Nigerian market, and in 2014 was voted the second best place to work in the country, showing the benefits that may be gained from TNCs. Shell is another company, helping to extract oil from the Niger delta; it has brought in a lot of tax and export money, employed 65 000 people and linked in with Nigerian industries. However, the oil operations are in a difficult area and there has been a long history of oil spills, which have contaminated the natural environment and created difficult living conditions for the Ogoni people.

Changing political and trade relationships

Nigeria is a member of the British Commonwealth, but is increasingly more involved in West Africa (Economic Community of West African States or ECOWAS trading group), African politics (African Union), oil politics (OPEC) and the UN (including being a member of the Security Council in 2014–15).

Nigeria now imports most goods from China, the European Union (EU) and the USA; a major import from China is cheap mobile phones and from the USA is petrol/diesel. It exports goods mostly to the EU, the USA and India. The main goods are raw materials such as oil (low in sulfur), cotton and rubber. Nigeria does not trade much with the rest of Africa at the moment; the only significant countries are Ghana and Côte d'Ivoire. In 2016, Nigeria had a negative trade balance.

International aid

Nigeria may be relatively wealthy, but it still has a large number of poor people. Consequently, it still receives international aid to tackle issues such as access to clean water, education and health care. As a member of the Commonwealth, Nigeria receives aid from the UK (health care in rural areas); as a trading partner it receives aid from the USA (children) and from the International Development Bank (medicines and medical supplies). Some aid is given by NGOs such as ActionAid, which has helped to set up health clinics to tackle disease.

Local community projects are more successful than large projects because the poor people can be targeted directly with what they need and corruption in government is avoided.

Environmental impacts

While many of the large industrial premises are subject to scrutiny, there are many small informal businesses, including illegal oil refining, in Nigeria. Consequently, there is a lot of land, air and water pollution near to urban and industrial areas. These are harmful to people and the natural environment (see Figure 11). Oil spills have polluted fishing areas and farmland, as well as the natural swampland, and oil companies have agreed to pay compensation to local people and to clear up the oil.

Figure 11 A recent oil spill in Nigeria

There has also been **deforestation** as farm and urban areas have expanded, reducing **biodiversity** and increasing land degradation.

Quality of life

Statistics show that there has been a slow improvement in the quality of life for the average Nigerian since 1990. Life expectancy has increased, the mortality rate has fallen, safe water access has increased and enrolment in secondary schools has nearly doubled. The HDI has increased slowly, from 0.466 in 2005 to 0.514 in 2014, but with no recent change in rank position (152nd). However, health staff numbers have levelled off and the proportion of the population with access to sanitation has fallen. One of the problems is the rapidly increasing population, which has nearly doubled since 1990, and another is that the oil wealth has not been used effectively. In 2015 the Terrorism Index for the country increased to an all-time high.

If the small improvements are due to the oil wealth, what will happen when the oil runs out? Nigeria will need to keep a stable government, help the areas of the country with environmental difficulties and increase the levels of trust between the different ethnic groups.

Economic sector	% of GDP
Services	42.64
Agriculture	21.97
Oil and gas (mining)	14.50
Information and communications	10.94
Manufacturing	6.83
Construction	3.12

Table 4 Nigeria: GDP share by economic sector, 2013

Development and change

There are eight areas covered above about the development of and change in Nigeria.

For each section, produce a revision card with four bullet points to learn ready for a nine-mark exam question. Make sure that you include some case study facts within your bullet points.

If you have studied a different country, do the same revision exercise, bringing together the information you have under the headings above.

Graphical skills

Appropriate graphical representation

Using Table 4, choose a suitable graphical method to present the data. Why is your choice the best one?

Advantages and disadvantages of TNCs

Make a list of the advantages and disadvantages of TNCs in Nigeria or the LIC or NEE you have studied.

CHECKIT!

1 a Give a definition of the term 'economic change'.

   b Give a definition of the term 'environmental change'.

2 For an LIC or NEE that you have studied:

   a List the advantages that the country has to help it develop economically.

   b Describe the cultural characteristics of the country.

   c Explain the socio-political challenges that exist in the country.

   d Explain the strength of influence on the country of political and economic globalisation.

3 To what extent has the quality of life changed in Nigeria, or the country you have studied, as the economy has developed?

Changes in the UK economy

Causes of economic change in the UK

Economic change means that the way of making money through businesses and industries has changed over time. The UK has experienced a change in industrial structure from primary industries before the industrial revolution, to secondary during the industrial revolution until about 1920, a decline of manufacturing especially after 1960 to be replaced by services and finally the growth of the quaternary sector since 1980.

Deindustrialisation and decline of traditional industries

1. In 1945, the UK still had an important world position in terms of manufacturing industries, but competition from the USA, Germany and Japan gradually reduced the UK's position. Then, from the 1950s, many former European colonies gained independence and started their own manufacturing industries in direct competition with the UK.

2. In the 1990s, several LICs became newly industrialised countries (NICs), such as South Korea, rapidly developing their manufacturing industries such as shipbuilding and car manufacture.

3. In the UK, coal mining declined rapidly in the 1980s due to cheaper imports and the increased use of oil and natural gas.

4. Manufacturing industries in the UK had old technology, were inefficient and often produced inferior products compared to Germany, Japan and South Korea, who were able to use the latest technologies.

Figure 12 Changes to the UK's industrial structure

Globalisation

Since the 1950s, the links between different places and countries around the world have multiplied.

- Modern transport, in the form of container ships and jet aircraft, has allowed goods (trade) and people (migration) to move easily from one place to another. With international ports such as Felixstowe and airports such as Heathrow, the UK has benefited and can export its specialist products more easily.

- Information technology, such as mobile technologies and the internet, has allowed instant worldwide communication for personal, entertainment and business purposes. This has enabled the UK tertiary and high-tech sectors to grow, such as London – as a world cultural and financial centre – and Cambridge Science Park.

- UK corporations have outsourced some parts of their business and jobs to cheaper English-speaking locations; this has caused some low-paid job losses and kept UK wages low.

- Foreign transnational corporations have invested in the UK, although they also compete with UK industries which cause closures.

- The UK has the greatest number of **diplomatic links** around the world and so is able to negotiate favourable trade deals. These political groupings are important in a competitive world.

Government policies

The UK government policy towards tertiary businesses and secondary industry has changed over time:

- At one time traditional industries, such as coal and steel, were **nationalised**. This means that they were run by the government to provide support and guarantee jobs, even if the industries were not profitable.

- Recently, UK government policy has been to privatise industries and services and to deregulate financial markets, although in times of economic crisis some nationalisation may still take place, such as with Lloyds TSB in 2008 (30 per cent public ownership).

- There is political disagreement over **privatisation**; benefits usually include greater efficiency and more cost effectiveness, better international competitiveness and the ability to attract foreign investment. However, negatives can be: loss of UK jobs, higher foreign ownership and profits leaving the country.

- Government investment in the road and rail system helps businesses to move things faster and more easily, saving money.

- The government has also encouraged FDI into areas of decline and established regional organisations to coordinate training and education of the workforce and encourage private business investment. This includes the establishment of business and science parks (i.e. modern industrial estates).

DO IT!

Industrial decline in the UK

Start, or add to, a new blog on your geography department's website that discusses the changes in industries and employment in your region of the UK since 1950.

STRETCH IT!

Economic recession

There is a cycle of major economic recession and boom. During a recession, all aspects of an economy are affected. Those businesses that are marginal, either because they cannot compete or are producing a product that is outdated, or are inefficient because their costs are too high, close down. Unemployment may then increase and people have less money to spend, which reduces the customers for other businesses, which then also suffer. The 2008 financial crash (caused by the over lending of money) brought a recession that economists following the theory of Kondratiev cycles expected. The UK is coping better with the recession than many other countries, as much of its outdated industries had already closed, the government supported the banking sector, and the UK is at the forefront of many modern industries and businesses.

The move towards a post-industrial economy in the UK

The UK is no longer a world centre of manufacturing industry. The **post-industrial** phase is being dominated by the tertiary and quaternary sectors of the economy. Often, these are in different locations from the traditional secondary industry areas, being based mostly in motorway corridors radiating from London rather than in the North-East, South Wales or the North-West.

Information technology

Information has become more easily accessible due to technology based on computers and satellite communications. In a relatively short space of time, the mobile phone and more powerful portable computers have completely changed the way people live and work. The UK is at the forefront of this technology.

1. There are lots of office jobs in the UK with about 1.3 million people working in the information technology (IT) sector.

2. People are able to work more flexible working hours and in different locations, such as working from home.

3. Businesses and work are no longer pulled towards sources of raw materials or transport hubs, so they are free to locate anywhere there is electricity and broadband.

4. Traditional industrial regions have declined.

5. London, being the capital with many national and international business links, has expanded; some regional cities have also expanded, making the UK one of the biggest digital economies in the world.

Service industries

Service industries are wide ranging, from health and education through to shops and insurance. As these have increased (now creating 79 per cent of UK economic output), so has the need for an educated skilled workforce.

- Many UK places have employment dominated by the **service sector**.

- Services are found everywhere, so these types of job are everywhere (although greatest where there are lots of people who need or use the services).

Finance

Finance mainly involves the banks and their related activities, including foreign currency dealings and stocks and shares. As people have become wealthier and the UK economy has remained relatively strong, even during economic recessions (such as 2008), there are lots of financial transactions and requirements.

- London is an international financial centre, with many banks attracted by business-friendly regulations and worldwide links.

- Two million people are employed in this sector and it makes an estimated contribution to the UK's GDP of 10 per cent.

DO IT!

IT benefits

Plan an answer to a question that asks you about the role of IT as the UK moves to a post-industrial economy.

Research

For all types of businesses to succeed in the long term, they need to carry out research and development (R&D). The UK has had a strong tradition of innovation and developing technology. The next wave of economic development will arise from current research into areas such as biotechnology, nanotechnology, medical, environmentally friendly and information services.

- Around 60 000 people are employed in research, with an estimated indirect contribution to the UK economy of £3 billion.

- Much of the research is located in or near universities, for example Norwich, in science or research parks, but also in other places around the world, such as the work of the British Antarctic Survey.

Science and business parks

All large urban areas have tried to establish modern business areas in accessible locations (near motorways and airports) – usually in the outer suburbs or **rural–urban fringe** where the surroundings are more attractive to work in (see Figure 13). These are designed to provide the latest buildings and communications technology in a small area where certain facilities and services can be shared by large and small companies.

- Being close to a good university, such as Cambridge or Edinburgh, also allows the businesses to get a highly skilled workforce and to work together on research.

- Future jobs and business success for a region may depend on the success of science and business parks to develop new ideas and also build the reputation of the region, for example, Cobalt Business Park in Newcastle on Tyne.

DO IT!

Features of a science park

1 Make a list of the key location factors that you would expect to find influencing a UK science park.

2 Where is your nearest science park?

Access	At the A14/A10 junction. The A14 links East and West. M11 and inter-city rail links to Stansted airport and London. Deliveries, workers and business meetings.
Workforce	Cambridge Science Park founded by Trinity College in 1970. Skilled workers straight from university; help R&D (science for biotechnology, geography for GIS).
Space	Inside A14 bypass avoiding green belt restrictions. About 70 companies. Avoids traffic congestion and has cheaper land to spread out on.
Problems	51 000 daily commuters, most by road. Narrow roads in city centre. Frequent congestion. People attracted and house prices consequently high.
Attractive	Historic city, nice countryside and villages attract skilled workers to live and work. Worldwide reputation for science, technology and academic areas.
Council support	Planning permission given to extend the science park and for new settlements (Cambourne 2008–16) built near Cambridge. County council developed Castle Park R&D site.

Figure 13 Location of Cambridge Science Park

Impacts of industry on the physical environment

Traditional industries created much visible pollution of the air, water and land, which resulted from burning fuels and using chemicals. Pollution from modern industry that uses electricity is not always visible at the sites. However, thermal power stations using **fossil fuels** to generate electricity contribute greatly to global warming, and resources must still be obtained from the ground, leaving scars on the landscape.

Environmental sustainability

Environmental sustainability means looking after the physical environment so that future generations can get the same benefits from the planet as current and past generations. **Green technology** is now available that can dramatically reduce pollution and damage to ecosystems and the landscape. Environmental laws are much stricter now, which makes companies take action to reduce their impact on the physical environment.

Case study

Making industry more environmentally sustainable: Torr Quarry, Somerset

Quarrying is common across the UK to supply rock, sand or gravel for construction. This involves digging a large hole in the landscape, creating a scar and changing vegetation and run-off patterns, as well as the creation of dust and noise by machinery.

Torr Quarry is a limestone quarry near Shepton Mallet on a 2.5 km² site that has been operating for over 70 years and supplies rock chippings for roads (previously known as Merehead Quarry). Eight million tonnes are produced a year and transported mostly by rail, which keeps heavy lorries off local roads and reduces air and noise pollution. Somerset County Council and the Environment Agency monitor operations and Aggregate Industries who run the quarry have produced a biodiversity management plan.

In 2009, plans were submitted (approved in 2012) to deepen the quarry and extend its life to 2040. At the same time, restoration plans were outlined. These restoration plans include:

* creating recreational lakes by flooding the quarry
* using the lakes as reservoirs for water supply
* creating new woodland areas and calcareous grasslands on landscaped natural rock areas
* retaining the railway link to help establish businesses and keep cars off local roads.

Ordnance Survey (OS) mapwork skills

OS mapwork 1:25 000 scale

Study Figure 14 showing a 1:25 000 scale map of the area around Torr Quarry. Remember that on this scale of map 1 cm is equal to 250 m.

1. Describe the types of economic activity shown on the map extract.
2. What type of vegetation is found in Monk Wood (6943)?
3. Apart from the name, what is the map evidence that there is an active rock quarry centred on 695446?
4. Give the six-figure grid reference for Lodge Hill Farm, found in the north-east of the map extract.
5. What is the distance (in km) by footpath from the underpass at 690439 to where the footpath crosses the railway line at 684431?
6. Describe the characteristics of the slope in grid square 6744.
7. Draw an accurate cross section to show the landscape from 680460 to 710430. Label it fully with information from the map.

Figure 14 Ordnance Survey map of Torr Quarry, Somerset

NAILIT!

OS maps

There are two scales of OS map that you need to know – 1:25000 and 1:50000. Make sure you know:

- What these scales mean in terms of 1 cm on the map, and how to measure straight and curved distances.

- How to interpret contour lines and spot heights to work out what a landscape is like.

- Some map symbols: this may help to save you time in an exam.

- How to use four-figure and six-figure grid references accurately.

- How to draw or interpret a cross section through the landscape using the contour lines.

DOIT!

Environmental sustainability

Do you think the actions of Aggregate Industries are enough to achieve environmental sustainability at Torr Quarry? Note down your answer, along with the reasons why you think this is the case.

131

Social and economic change in the UK rural landscape

- **Commercial farming** still dominates large parts of the countryside with its large fields, machinery and buildings, but the number of people working on farms has greatly decreased. Farms have had to diversify into other ways of making money, such as campsites, farm shops or workshops for small businesses. In more remote rural areas, there has been **depopulation** as young adults leave to look for alternative and higher-paid work in cities. Some of these areas have attracted seasonal tourists and second-home owners, but this has not been enough to stop services closing, such as village shops, primary schools and post offices.

- Accessible rural areas near large towns and cities have experienced an influx of people who work in the urban areas but seek cheaper housing, the quieter home life offered by rural areas and the recreational opportunities provided in rural areas. With modern IT, people are also able to work from their rural homes. Rural areas of southern England are also retirement migration destinations. These movements have increased pressures on the countryside, often urbanising it.

Case study

A UK rural area with population growth: South Cambridgeshire

South Cambridgeshire is an accessible rural area with access to London via the M11 and intercity rail, as well as the university city of Cambridge itself. There are many picturesque villages and open farmland, but also new settlements and expanded settlements. The population of the area is currently around 150 000 and is predicted to reach over 180 000 by 2031 due to migration for work from within the UK and abroad.

The impact of population growth

- Commuters tend to use urban services rather than support rural ones.

- Increased traffic on small rural roads causes congestion and increases the risk of accidents.

- Community changes as population background changes from traditional rural (farming) to exurbanite.

- House prices increase dramatically due to demand and shortage, so young adults cannot afford housing and so move away, leaving behind an ageing population.

- Uncertainty in farming and farm employment due to urban pressures on land; farmland lost to urban uses.

- Prices in shops are above average due to wealthy customers and demand.

- Migrants from Eastern Europe may put pressure on services and housing stock, but provide a cheap workforce for farmers and food-processing industries.

DO IT!

UK rural area with population growth

Make a revision card showing four summary bullet points outlining the impact of population growth in a UK rural area.

Case study

A UK rural area with population decline: the Outer Hebrides

The Outer Hebrides are a chain of islands 210 km long, 65 km off the western coast of Scotland. They are very remote, being about 470 km from Glasgow in Scotland and nearly 1130 km from London. They have a population of 27 400 people and only one town, Stornoway, on the main island of Lewis. They are exposed to gales from the Atlantic Ocean, and farming is marginal and based on crofts. The population is in decline due to people leaving, especially young adults, for the greater opportunities on the mainland. A few Syrian migrants have been relocated to the area.

Impact of population decline

- There is an ageing population (the majority are currently between 45 and 74) and older people need specialist services.

- The number of children is declining, so schools face closure.

- There are fewer people of working age to do the jobs that need doing on the islands.

- Services are closing or being faced with closure or restricted services, for example, ferries which are a vital link to the mainland.

- Farming and fishing jobs have declined, and fish farming has not developed in the area.

- Tourism has increased, but at the moment there are not enough facilities to cater for the increases needed to sustain the economy of the area.

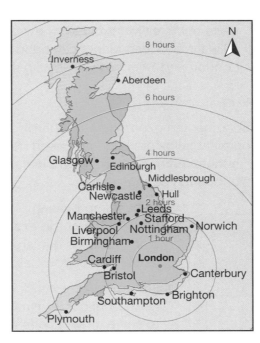

Figure 15 Isoline map of distances/times by train from London (on the right-hand map, a distorted UK shape is included to show the effect of travel times).

DO IT!

UK area with population decline

Make a revision card showing four summary bullet points outlining the impact of population decline in a UK rural area.

Transport improvements and developments

Transport links have always determined accessibility, and places with better access to workers, raw materials, energy supplies, water, components or markets are attractive to industries and businesses. Improvements in transport have always been seen as a way of encouraging economic development in an area or region.

Road infrastructure

The UK has an extensive motorway system. However, some regions are poorly served, such as East Anglia and the South-West, and many are at or near capacity. Motorways and A-roads help growth of businesses in many large urban centres. Fast, direct roads help provide access to markets for businesses regardless of where they are located in the UK, and motorways provide corridors along which modern industries locate.

- Recent developments include the M6 toll motorway, bypasses and the Dart Charge (replacing toll booths at the Dartford Crossing).
- Future developments include adding extra lanes to motorways and widening the A303 to ease congestion on a key route to the South-West.

Rail infrastructure

The UK has an ageing railway network that is inefficient in some locations and congested in others. As oil runs out, the electrified railways will become more important to provide for commuters, enable business meetings to take place in key cities across the country and help take vehicles off congested roads.

- Recent developments include High Speed 1 (HS1) linking London through Kent and the Channel Tunnel to the European mainland, and minor improvements to freight routes linked to container ports such as Felixstowe.
- Future developments include High Speed 2 (HS2), planned to link some northern cities with the economic core of the South-East, cutting travel times dramatically. Crossrail in London will also link western and eastern parts of the city to provide a new commuter route. Electrification of railways between Manchester and York, Liverpool and Newcastle, and London and Sheffield will all help the northern cities.

Port capacity

Importing and exporting from the UK is mostly done through ships. Modern container ships are huge and need deep water and specialised cranes to deal with them. Ferries are important in carrying people and freight to and from neighbouring countries.

- Recent improvements include main ports such as Felixstowe and Southampton expanding their docks and making further plans as ships get larger. New port facilities on the Thames estuary – London Gateway – have also been developed.
- Future developments include the completion of Liverpool2 (a new container terminal) and investment in Belfast for facilities to serve growing offshore energy technologies.

Figure 16 The world's largest container ship at the UK's largest container port, Felixstowe

DOIT!

Transport improvements

Discuss with your parents and a friend what transport improvements are needed in your region of the country. You need to be able to answer this question: Why are these improvements needed?

SNAPIT!

Port capacity

Snap an image of Figure 16, showing a container ship at a UK port. Use this image to remind yourself about the benefits for employment and economic development of such a large modern port facility. Perhaps annotate the photo.

Airport capacity

All UK regions are served by regional airports with several now having international airports employing 300 000 people. The UK is linked to 114 countries by air, and there are 420 000 flights within the UK each year. However, London is regarded as having reached its airport capacity and there is much debate about how to increase it.

- Recent developments include consolidating and expanding the roles of regional airports, many of which now handle international flights. Proposals to expand Stansted have proved controversial, but the airport has expanded its passenger capacity.

- Future developments may include a new airport in the Thames estuary area to the east of London – a relatively poor part of the region, which is also being improved through the Thames Gateway initiative. There has been much debate over the 2016 government decision to increase capacity by building a new runway at Heathrow – with Gatwick airport claiming that it is still the best option.

Practical skills

Heathrow–Gatwick debate

Read the points made in Figures 17a and b. Produce a summary argument to present to the government in favour of building a new runway at one of these airports. In this summary, you should give the main points for your choice and the main reasons for rejecting the other.

Figure 17a Publicity issued by Gatwick airport

If we want Britain's economy to keep growing, we need to grow Heathrow.

It is no surprise that Britain's ability to achieve sustained economic growth depends on our ability to trade with other countries.

For an island nation, it always has and it always will.

But the countries we need to trade with are changing, fast.

By 2050, nearly half of global GDP will be derived from emerging markets: primarily in Asia and South America.

Britain's exporters desperately need direct access to all these burgeoning markets.

And only one British airport can provide it.

Heathrow is one of a handful of hub airports around the world with world-class connections, infrastructure, transport links and goods-handling skills. It enables British companies to compete for their share of this huge international prize.

What we don't have is the capacity to expand. And we need it.

Every month this problem goes unsolved costs the British economy £1.25 billion through lost trade.

While Britain wrings its hands over the site of a new runway, our competitors in France, Germany and The Netherlands are rubbing theirs.

They have the hub capacity. They have the will to grow it. And they have the wholehearted government support to take what could be Britain's.

A new runway at Heathrow would create more than 120,000 new jobs.

But that 3,500 metre stretch of tarmac would also deliver at least £100 billion of economic benefits: the length and breadth of Britain.

In new orders that fuel UK businesses, large and small. In profits and dividends that feed our pensions and ISAs. And in taxes that fund our public services.

We are ready to help Britain compete in the global race for growth and jobs: because it's what we do already.

But our ability to support British economic growth in the future depends on Britain's willingness to let us expand now.

This issue has been in the long grass for long enough.

If we want Britain to get on, let's get on with it.

Heathrow
—Taking Britain further—

heathrow.com/takingbritainfurther   #TakingBritainFurther

Sources: Goldman Sachs Global Investment Research, 2012; Frontier Economics, "Missing trade opportunities: The impact of Heathrow's capacity constraint on the UK economy," November 2012; Frontier Economics, "Employment impacts from growth at Heathrow," April 2014.3

Figure 17b Publicity issued by Heathrow airport

The UK North-South divide

- Many northern regions of the UK had a dependence on manufacturing industry and have suffered greatly due to **deindustrialisation**, with its consequent high unemployment and lower house values.

- For many decades, there has been a divide in the UK, with the peripheral areas of the North and West being generally poorer than the core area of the South-East.

- A wide range of indicators (social and economic) all show that there is a North-South divide within the country, with the North being generally poorer than the South, although the North has some richer areas and the South some poorer areas.

DO IT!

North to south movement

Draw a flow diagram to show why, for several decades, people in the UK moved from northern regions to southern regions.

- There has been a southward drift of population seeking job opportunities, although this appears to be slowing, with more migration between northerly regions recently.

- Southern regions have had growth in the tertiary sector and, therefore, have experienced lower unemployment levels but much higher house prices.

Strategies to reduce differences

Governments provide funds to rural areas that have few people and few jobs, and to former industrial areas that have experienced deindustrialisation and job losses.

1. Until the UK leaves the EU, the European Regional Development Fund (ERDF) provides money for regeneration projects and better communications such as broadband, for example. (See Figure 18.)

2. Better transport is developed, such as improved rail links between Manchester and Sheffield and HS2 between London and Manchester.

3. UK government grants and tax incentives are offered through Enterprise Zones, and 24 new ones have been created since 2011. The government has also set up Local Enterprise Partnerships (LEPs) since 2011. These aim to bring together local authorities and businesses to identify opportunities for job creation. (See Figure 18.)

4. FDI is made and other businesses are located in declining areas, such as the Nissan car factory in the North East.

UK assisted areas, 2014–20
- sparsely populated areas
- areas with lower than average GDP
- other disadvantaged areas

0 — 700 km

Figure 18 UK assisted areas, 2014–20. Assisted areas are places where regional aid can be offered to businesses, with the aim of helping disadvantaged regions

The UK in the wider world

Since the days of the British Empire, the UK has had a strong influence on the rest of the world. With globalisation and other countries gaining influence, the UK influence has declined but it is still relatively strong. Migration between Commonwealth countries and the UK has maintained these historical links, and membership of the EU created closer economic and cultural links with other European countries. The UK is also a member of major organisations such as the G8 (economic), NATO (security) and the UN Security Council.

Trade

While exports of manufactured products have declined due to deindustrialisation, the UK still sells many key products such as machinery and transport equipment. It also imports goods from all around the

NAILIT!

North-South divide

Remember that it is difficult to draw an exact line between a 'rich' South and a 'poor' North in the UK because there are richer and poorer areas across the whole of the UK. Some may draw a line from the Severn estuary to the mouth of the Humber, but then the average salary in the North-West is higher than the East Midlands, and recent migration trends appear to have shifted from being North to South to being more between the northern regions.

world. Most UK trade is with EU countries overall, with exports highest to the USA and Germany; imports are highest from Germany and China.

Culture

English is an international language and the main language in many countries formerly colonised by the UK. Along with this, the legal system, sport, affiliation to the monarch, UK music, TV and film are major parts of the culture of many Commonwealth countries around the world. Many students take UK-based examinations or attend UK universities because of the high academic reputation that these have. The UK is a multicultural society with a variety of cultures introduced from Commonwealth countries at different times, such as India and, more recently, from the EU.

Transport

The UK has major international ports at Felixstowe, Southampton and the Channel Tunnel, and major international airports at Heathrow, Gatwick and Glasgow, which link to other hubs around the world. In 2015, Heathrow was linked to 84 countries, served over 75 million passengers and conveyed 1.5 million tonnes of freight. In 2014, Felixstowe was linked to 400 other ports and received 3000 ships (see Figure 16); 42 per cent of the port traffic was with the EU – especially the Netherlands. In 2015, the Channel Tunnel carried around 21 million tonnes of freight and about 21 million people.

Electronic communication

The internet has allowed a rapid increase in the ability to connect to any part of the world, and internet businesses have grown in the UK. News is instant and UK businesses are able to access new markets anywhere in the world. The UK is a focal point for submarine fibre cables with connections to North America and Europe, and it is planned to lay a cable from the UK to Japan via the now ice-free Arctic Ocean, north of Canada.

Economic and political links

The UK has some traditionally strong economic and political links with:

- The USA, with trade across the North Atlantic and membership of NATO based on common language and economic (free trade) and political aims.

- The Commonwealth in different continents, with support through preferential aid (Department for International Development – DFID) and trade that helps each country. Trade and cooperation takes place between these countries, as well as the promotion of democracy, human rights and economic development. The Commonwealth Games are also a major world sporting event.

- The EU, as they are the UK's closest neighbours and have worked collectively to secure European jobs in the face of international competition and on common world political aims. Until the completion of the UK's exit from the EU, various EU funds, such as URBAN, ERDF and Common Agricultural Policy (CAP) have supported poorer areas of the UK, while in the EU workers can move across borders and this has helped the UK in farming, food processing and construction.

CHECKIT!

Causes of economic change in the UK

1 Name the UK's main economic sector in the 21st century.

2 Give two reasons why deindustrialisation took place in the UK.

3 Describe one government policy that tries to help the UK economy.

4 Explain how globalisation may help the UK's economy to be successful.

The post-industrial economy of the UK

5 Explain one way in which a motorway helps an industry or business.

6 Describe how IT may help to improve the UK economy.

7 Give two common location factors for UK science parks.

8 Explain why research and development is important to the future of UK businesses and industries.

Impacts of industry on the physical environment

9 Name one way in which an industry may damage the physical environment.

10 Explain how using 'green technology' may reduce impacts on the physical environment.

11 For an example of environmental sustainability that you have studied:

   a Describe the problems created by the economic activity.

   b Explain how successful the actions to reduce the impacts have been or will be.

12 a Give one change that has taken place in remote rural areas of the UK.

   b Give one change that has taken place in accessible rural areas of the UK.

13 For a UK rural area that you have studied that has experienced population growth, describe two impacts on the rural area caused by the population growth.

14 For a UK rural area that you have studied that has experienced population decline, explain one impact on the rural area caused by the population decline.

Transport improvements and developments

15 a Name one recent improvement to the road infrastructure in the UK.

   b Describe one possible future improvement to the UK rail infrastructure.

16 a Describe a recent improvement to port capacity in the UK.

   b Explain why there is debate about where to expand London's airport capacity.

The UK North-South divide

17 Give one significant difference between the North and South of the UK.

18 a Name one UK government scheme to help reduce the North-South divide in the UK.

   b Explain how improvements to the provision of broadband in northern areas of the UK would help reduce the North-South divide.

19 Name one major international organisation that the UK is a key member of.

20 Describe the UK's current links with the EU.

21 Explain why transport hubs such as ports and airports are important to maintaining the UK's links with the wider world.

The changing economic world

Economic development and quality of life

1 Outline four ways of measuring quality of life.

2 a Name two LICs.

 b Name two NEEs.

 c Explain the differences between an LIC and an HIC.

3 a Give a definition of the term 'gross national income'.

 b Explain how data on infant mortality can be used to measure development.

 c Explain the strengths and weaknesses of using HDI to measure development.

4 a Describe the trends of the birth rate and death rate at Stage 3 of the DTM.

 b Explain how the changes to the birth rate in Stage 3 of the DTM are linked to the development of a country.

5 a Explain how climate may be a cause of uneven development around the world.

 b Explain how international trade may lead to uneven development around the world.

 c Explain how development aid may be responsible for some of the uneven development around the world.

6 a Describe the wealth disparity in one country or world region.

 b Give one reason for the health disparities in LICs.

 c Explain how a cycle of poverty in an LIC keeps people poor.

7 a Give a definition of the term 'refugee'.

 b Explain the push and pull factors involved with international migration from an LIC to an HIC.

Global development gap

1 a Give a definition of the term 'remittances'.

 b Give a definition of the term 'infrastructure'.

 c Explain how improvements to infrastructure in a country can help it to reduce the development gap.

 d Explain why a multiplier effect can be considered to be self-sustaining.

2 a Give a definition of the term NGO.

 b Describe the types of development project that an NGO usually supports.

 c Explain why some international aid may not help to reduce the development gap.

3 Give two reasons why intermediate technology should be used in development projects.

4 Explain why fairtrade may be better at reducing the development gap than normal trade between countries.

5 a Name two schemes for reducing the debts of LICs or NEEs.

 b Explain why microfinance loans from local banks to people in LICs may be better for reducing the development gap than loans to a country from the World Bank.

6 For a country that you have studied that is using tourism to reduce its development gap:

 a Describe the state of the tourist economy in the country.

 b Explain the benefits brought by tourism to the country.

 c Explain the problems linked to tourism that the country experiences.

Rapid economic development and change

1 a Give a definition of the term 'cultural change'.

 b Outline the differences between economic change and social change.

2 For an NEE that has experienced rapid economic development and change:

 a Give two advantages that the country has to help it develop economically.

 b Outline two obstacles that the country faced in the early stage of its development.

3 For an NEE that has experienced rapid economic development and change:

 a Describe one political change that has taken place as part of the country's development.

 b Describe one environmental change that has taken place as part of the country's development.

 c Describe one cultural change that has taken place as part of the country's development.

d Describe one social change that has taken place as part of the country's development.

e Suggest whether these changes have been positive or negative. Explain your answer.

4 For an NEE that has experienced rapid economic development and change:

a Describe how the country's industrial structure has changed.

b Explain whether foreign TNCs have been a benefit or a problem for the country.

c Describe the country's political and trading links with the rest of the world.

d Explain why the country still needs to receive international aid.

e Explain why the quality of life in the country has not increased as rapidly as expected.

Changes in the UK economy

1 **a** Give a definition of the term 'quaternary sector'.

b Give a definition of the term 'deindustrialisation'.

2 Explain why the UK experienced deindustrialisation.

3 Explain how globalisation has helped the UK economy.

4 **a** Give a definition of the term 'privatisation'.

b Explain how FDI can help the UK economy.

c Compare the advantages and disadvantages of nationalisation and privatisation.

5 **a** Explain how IT is helping to develop the post-industrial economy of the UK.

b Explain how service industries and the financial sector are helping to develop the post-industrial economy of the UK.

c Explain why a large proportion of the UK's research and development is found in science parks.

6 Give a definition of the term 'environmental sustainability'.

7 For an example of environmental sustainability that you have studied:

a Outline how the economic activity caused damage to the landscape and natural environment.

b Explain how the restoration plans for this location have the potential to achieve environmental sustainability.

8 **a** Give a definition of the term 'depopulation'.

b Give a definition of the term 'retirement migration'.

9 For a UK rural area with population growth that you have studied:

a Describe the accessibility of the rural area.

b Explain four impacts of the population growth on the rural area.

10 For a UK rural area with population decline that you have studied:

a Describe the accessibility of the rural area.

b Explain four impacts of the population decline on the rural area.

11 Describe how road and rail improvements are improving accessibility to peripheral and poorer areas of the UK.

12 Explain why it is necessary to improve the port and airport capacities of the UK.

13 **a** Name a northern region of the UK.

b Give three differences, in terms of human geography, between the north and south of the UK.

c Explain two strategies for reducing the differences between the north and south of the UK.

14 **a** What is meant by the term 'Commonwealth countries'?

b Name two other major international organisations that the UK has a significant role in.

c Compare the roles of trade and culture in increasing the UK's influence in the world.

d Explain why the UK is a focal point for world transport and electronic communications.

e Explain why it is important for the UK economy to keep strong political links around the world.

The challenge of resource management

THE EXAM!

- This section is tested in Paper 2 Section C.
- You must know Resource management.
- You must know **one** of food, water or energy (you do not need to revise more than one).

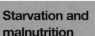

NAIL IT!

Starvation and malnutrition

Remember the difference between these terms. 'Starvation' is the result of not enough food – the World Health Organisation (WHO) says that people need at least 2000 calories a day. 'Malnutrition' (or undernutrition) is the result of an unbalanced diet that does not provide the human body with the vitamins and minerals it needs to stay healthy and fight disease (see Figure 1). Some people may be obese. This is where they have eaten too much food or have not had enough exercise to use up calories. Obesity can also lead to health problems.

Resource management

The significance of food, water and energy to economic and social well-being

Food, water and energy are fundamental to human development. The overall development of countries and the **quality of life** for people depend on a number of **resources** being readily available. A resource is something that can be used by people.

Food

Food is an essential of life. Without enough of it people would die of **starvation** and without enough variety in their diet they would be **malnourished** and unable to fight disease. Food provides calories which give the human body energy to develop physically and enable people to become stronger workers. **Food security** becomes an issue when prices increase rapidly.

STRETCH IT!

FAO

The Food and Agriculture Organization of the United Nations (FAO) reported that the number of undernourished people declined between 2005 and 2015, but that 800 million people still remained hungry (see Figure 1).

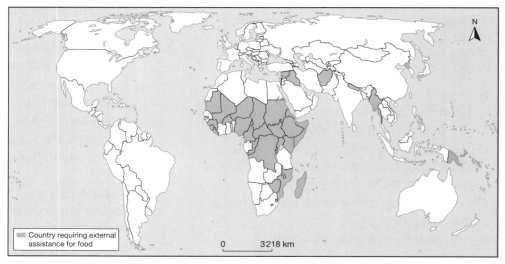

Country requiring external assistance for food

0   3218 km

Figure 1 Countries requiring external assistance with their food supply (June 2016)

Water

Water is an essential of life. Without clean fresh water people cannot stay healthy for long. Water is also needed by industries, either in the processing of products or to cool machinery and for farming. These uses can conflict when water is polluted by industrial or agricultural wastes and water demands are increasing to a point where there is not enough to go around.

Energy

For people and industries to progress beyond a basic level, a source of energy is needed. Within their daily lives, people need energy for cooking, heating and light, for example. Industries and businesses need to have enough energy to operate on a larger, more efficient scale. There is a range of energy resources available from non-renewables like oil to renewables like hydro-electric power.

Global inequalities in supply and consumption

Food

Supply is influenced by:

1. **Physical factors** such as climate and soil fertility: warmer, wetter areas with fertile soils, such as the UK, will produce more food than drier, colder areas with infertile soils, such as northern Canada. Access to the sea for fisheries is important in coastal countries, such as Indonesia, for food supply.

2. **Human factors** such as population size and growth, level of technology, market networks and transport systems. Areas with a smaller population will have a larger supply of food per person than areas with a large population, although trading networks can move food supplies from one area of the world to another, such as the export of wheat by the EU, rice by India, fish by Chile and meat by the USA.

Wheat production (top five countries)		Rice production (top five countries)	
Country/world region	Million tonnes	Country/world region	Million tonnes
EU	154	China	143.4
China	129	India	105.6
India	89	Indonesia	45.1
Russia	62.5	Bangladesh	34.8
USA	54.4	Vietnam	28.9

Table 1 Estimated production of staple crops, 2016

DO IT!

Summarise why water, food and energy are considered to be essential resources.

STRETCH IT!

Food and Agriculture Organization of the United Nations (FAO)

In June 2016, the FAO reported that 37 countries needed outside help with their food supply. Of these countries, 28 were in Africa.

DO IT!

Increasing food consumption

Draw a diagram to show as many reasons as you can think of why food consumption increases as low income countries (LICs) turn into newly industrialised countries (NICs) or newly emerging economies (NEEs).

Graphical skills

Pictogram

Use the data from Table 1 for **rice** production to produce a pictogram to show the five countries:

1. Use Excel or similar software to construct the pictogram. Put millions of tonnes on the vertical axis and the five countries along the horizontal axis. The vertical axis should go from 0 to 150.

2. Select bar chart and adjust the width of the bars so that they are wide yet do not quite touch.

3. Select each bar in turn and select the fill option of 'insert picture'. Insert a picture of the flag of each country. (You may need to download a picture of each from the internet before you start this exercise.)

4. Fully label the axes and add a title.

Pie chart

Use the data from Table 1 for **wheat** production to produce a pie chart to show the five countries or world regions:

1. Add the five wheat production totals.

2. Take each million tonnes total for each country (or world region) in turn and multiply it by 360 (the number of degrees in a circle) and divide by the total millions of tonnes calculated in step 1.

3. These calculations give you the number of degrees for each slice of the pie chart. Draw the pie chart using a pair of compasses, protractor, ruler and pencil.

4. Fully label each slice of the pie chart and add a title.

Consumption is influenced by:

1. **Population numbers**: a larger population will require more food, and the world population is predicted to keep increasing at least until the middle of this century.

2. **Wealth of the population**: with more money, the demand for food increases. Therefore, as countries develop economically, demand will increase.

3. **Changing diet**: when people have more money, diet moves away from traditional local foods towards imports or different foods and, therefore, a wider choice.

Water

Supply is influenced by:

1. **Physical factors**: climate determines where rain falls. Mountain ranges have higher amounts of rainfall and rivers form on their slopes. Countries such as Germany have plentiful water, while Saudi Arabia and the United Arab Emirates (UAE) do not.

2. **Human factors**: the ability to build dams and reservoirs, extract water from the ground and build water treatment works is important. Those areas with the money and technology, such as France, provide their population with water to their homes, while those without water infrastructure, such as Sudan, do not.

Consumption is influenced by:

1. **Population numbers**: a larger population will require more water to drink and use in homes. The world population is predicted to keep increasing up to 2050, especially in Asia and sub-Saharan Africa. Higher densities of population also increase demand locally, such as in cities like Bangkok or Hyderabad (see Figure 2).

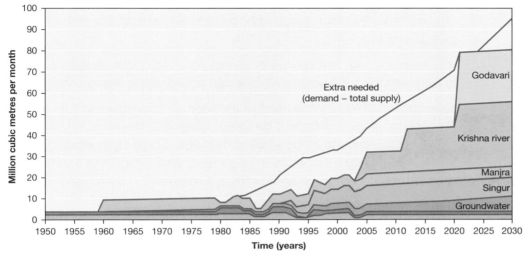

Figure 2 Hyderabad, India: water demand and supply pattern (all names refer to rivers)

2. **Economic development**: with many countries now developing economically, water consumption is increasing in countries such as China, India, Brazil, Mexico and Indonesia, because there is an increased demand from people, industries, energy production and farming. Farming is the biggest user of water in any country.

Energy

Supply – there is an uneven distribution of energy sources around the world:

- **Fossil fuels** are located where **geological processes** formed them millions of years ago, so oil is concentrated in certain regions like the Middle East.

- **Geothermal energy** can be obtained where there is **magma** close to the ground surface, so this is concentrated in a few countries such as Iceland.

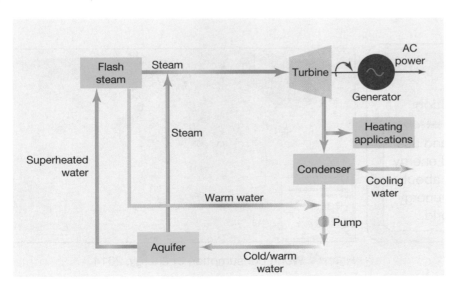

Figure 3 Geothermal power system

STRETCHIT!

Water footprint

The USA uses twice as much water as the global average (consumption per person), while Bangladesh uses just a third of the global average. The Water Footprint Network estimated global average water consumption to be 1385 m³ per person per year between 1996 and 2005.

DOIT!

Climate change and water

Make a list of the ways in which **climate change** will affect the two supply factors and the two consumption factors causing water inequality.

- Where there are steep-sided valleys and plentiful rainfall, countries can build dams and store water in reservoirs to release to create hydro-electric power (HEP) (e.g. Three Gorges Dam, China), or large rivers can be used with a system called 'run of the river', which just uses the natural flow levels of a river (e.g. Santo Antonio, Brazil).

- Solar energy can be used where there are clearer skies to receive long hours of, or intense, sunlight (e.g. Germany).

- Wind energy requires open areas exposed to fairly strong consistent winds (e.g. Denmark).

- Biomass, mainly wood, is a source of energy for many poor people of the world as they neither have access to, nor can afford, other energy resources.

These supplies are affected by the level of technology of a country: for example, the energy resource cannot be used unless the country can extract the fossil fuel from the ground or build wind turbines. Some resources are in remote locations, making them difficult to extract, and remote rural areas in developing countries do not have an electricity supply.

Consumption is influenced by:

- The demand for energy has been increasing over time. This is because more and more countries, for example China, are emerging economically and their industries and businesses use more energy. Also people have become wealthier, which has increased the household use of energy, especially transport.

- Fossil fuels (oil, coal, natural gas) are still the major source of energy around the world. But supplies are finite and will run out.

- Since 1990, the world's energy consumption has increased by over 60 per cent. Developed countries use seven times more energy than NEEs and 14 times more than LICs.

- Top consumers of energy per person are Canada, the USA, Iceland, Norway, Sweden, Finland, Russia, Saudi Arabia, Bahrain, the United Arab Emirates, Australia and New Zealand. Those using the least are in sub-Saharan Africa and South Asia.

- Sub-Saharan Africa depends mostly on biomass, with about 700 million people cooking on open fires.

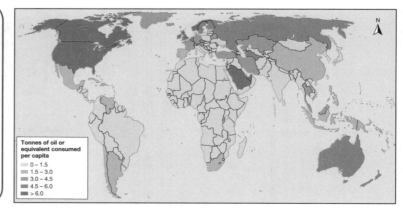

Figure 4 World consumption of energy, 2014

Opportunities and challenges in the UK due to changing demand and supply of resources

UK food

Changing demand

- Traditionally, people in the UK would have eaten food available according to season and produced by British farmers; some products came from Commonwealth countries and then, with membership of the EU, foods were easily available from other European countries. People have become used to a wider choice of food.

- With the expansion of supermarkets, and their ability to purchase in bulk and obtain foods from further away, consumer choice increased greatly and some UK produce was replaced by imports, such as apples from South Africa. Today, nearly half of the UK's food supply is imported, as consumers have become used to a greater choice and with increased wealth are able to buy more and a greater variety of foods. Seasons are now less important, as food products can be imported from the southern hemisphere and the tropics.

- There has been a move towards local farmers' markets and buying produce directly from UK farms, but these are relatively small scale compared with supermarkets. There have also been campaigns to 'Buy British'.

- Immigration to the UK has increased demand for foods from the home countries of migrants, and small shops and supermarkets stock world foods.

- Demand for organic products has increased due to their healthier and more environmentally friendly credentials with no use of artificial chemicals. However, these products are more expensive and so during times of economic hardship sales decline. Some people have an allotment or a vegetable patch in their garden.

Carbon footprint

- With a larger amount of imported food there is a cost to the environment. As imported food has to be transported to the UK, fuel is burnt by ships, lorries and aeroplanes, which contributes to carbon dioxide emissions and climate change. The term food miles is often used to remind people that the greater the distance the food has travelled, the higher the carbon dioxide emissions.

- Farming methods may also contribute to the carbon footprint, such as, the use of heavy machinery; drying grain in a silo; heating greenhouses; and freezing or cooling foods during transport.

- Imported food does not always have the larger carbon footprint, as transporting by ship is more energy efficient than other forms of transport, and some products grow outdoors in other countries, while in the UK they need heated greenhouses.

DO IT!

World foods

Visit a supermarket and note any special sections or aisles set aside for foods that are in demand from immigrant populations. Which countries or world areas are featured?

Carbon footprint

Draw and label (or annotate) a footprint outline diagram to show how UK food supply contributes to emissions of carbon dioxide into the atmosphere.

Agribusiness

- Traditionally, farming in the UK was done by families, often passing the farm from one generation to the next. However, this has been in decline for some time. Instead, there has been a growth in large intensive **commercial farms** because this lowers input costs and therefore increases profits, e.g. factory farming.

- A higher level of technology is used, including very large machinery, remote sensing with satellite technology and computerisation of chemical distribution on soils. These commercial farms are known as **agribusinesses** and have a lot of money invested in them by the companies that own them.

- Some of these farms may be owned by companies that also own the food-processing factories and have very strong links with supermarkets.

UK water

Changing demand

The amount of water used by people in the UK has rapidly increased in recent decades. This is due to:

- Homes having lots of modern appliances for washing things; constant access to hot water means that people shower more frequently.

- Demand for early seasonal foods, which encourages farmers to water their crops more.

- People spending more time on leisure activities such as gardening, swimming and playing golf, all of which use a lot of water.

- The population of the UK is increasing, although now at a fairly low rate.

- Although the UK is moving towards being a post-industrial country, it retains quite a few industries, many of which use water.

Quality

- Unfortunately, nearly three-quarters of the UK's surface water is below the 'good' level set by EU standards. Rivers provide fresh water, but may be polluted by farm chemicals, untreated industrial waste, spills from sewage treatment works and run-off from roads.

- Drinking water may be contaminated, fertilisers may lead to a process known as eutrophication (where algae consume oxygen and block sunlight), microbacteria from sewage may cause illness and disease, and people relying on rivers for an income (e.g. tourism) may lose out.

- There are laws to prevent pollution and water is treated before being piped to people, for example, in sewage works before being put back into a river and in water treatment plants. Roads and industrial premises have catchment traps to collect contaminated run-off.

Matching supply and demand

- The UK's **latitude** and position next to the Atlantic ensures that there is a **water surplus**. However, the east side, especially the South-East and East Anglia, only receives about a quarter of the rain that the western region gets.

- Unfortunately most of the UK population lives in the South-East and fewer people live in the wetter western areas. Therefore, the natural supplies do not match demand. **Water stress** or **water deficit** therefore exists from Kent to Hampshire and West Sussex to Cambridgeshire.

Transfer

- Transferring water around the UK has been an idea for some time, generally moving water from wetter areas to drier ones through pipelines and transfers between river systems. However, this is very expensive, so the long-existing plan to transfer water from the North-East all the way to London has not yet happened.

- Water is transferred from the Lake District in the North-West to big cities such as Manchester.

- Water is transferred from the large reservoir called Kielder Water in the North-East to the headwaters of the River Tyne, and from there to other rivers such as the River Tees, so linking the major cities of the North-East to reliable water supplies.

- The Yorkshire Grid Scheme transfers water to Leeds.

Figure 5 UK rainfall pattern and major rivers

SNAPIT!

Transferring water to the drier South-East

Snap an image of Figure 5, showing rainfall and rivers in the UK. Study this and add arrows and labels to show how you would transfer water to the South-East. Is long-distance transfer of water in the UK cost effective?

DOIT!

UK water supply and demand

Plan an answer to the question: What are the challenges of providing enough water to all parts of the UK?

STRETCHIT!

Water transfer costs

A 2006 Environment Agency study found that transferring water from the North-East to the South-East would cost four times as much as building new reservoirs in the South-East to meet the demand. The cost of the scheme was estimated to be about £42 million per mile.

UK energy

Changing energy mix

During the UK industrial revolution, coal was the most important energy source. Since the 1950s, we have been in the 'oil age'– although not much oil is used in the production of electricity. Energy demand had been increasing until recently, but greater efficiency of use, higher costs and conservation strategies have stabilised energy use in the UK.

The UK's primary energy mix is diversifying because:

* Oil is due to run out in the second half of this century.

* Natural gas supplies are uncertain as they have to be transported long distances by pipeline.

* There is concern over (and are agreements on) climate change as all the fossil fuels release carbon dioxide into the atmosphere.

* Renewable energy technologies have been developed.

Fossil fuels still dominate the UK energy mix, but the use of renewable energy sources, such as bioenergy and wind, is increasing.

Reduced domestic supply

1. Coal reserves in the UK have decreased and only the coal that is more difficult to obtain or would cause great environmental damage is left, making it more expensive than imported coal.

2. The UK has natural gas reserves in the southern part of the North Sea, but these have largely been used up (37 per cent left), and supplies have been imported since 2004.

3. Oil was obtained from under the central and northern parts of the North Sea, but these reserves are now significantly depleted (42 per cent left). Some new oilfields are still being opened (e.g. east of the Shetland Islands), but extraction peaked in 1999 and production is now only a third of peak production (see Figure 6 on page 152).

4. UK and EU regulations regarding carbon dioxide emissions are forcing the closure, or conversion to other fuels (e.g. biomass), of coal-fired power stations.

5. Nuclear power stations are reaching the end of their productive stage and radioactivity is still worrying, especially after disasters such as at Fukushima, Japan, in 2011.

6. There has been a large investment in wind energy, especially offshore wind farms, and fields of solar panels (renewables are up over 400 per cent since 2000). However, these only generate a small amount of the UK's electricity and funding is still uncertain.

7. Fracking for gas in shale rocks has also been suggested, but this is controversial because of environmental side effects, such as minor earthquakes, ground subsidence and chemicals polluting groundwater.

Primary fuel	1990	Index	2000	Index	2010	Index	2014	Index
Oil	100.1	100	138.3	138.2	69.0	68.9	43.7	43.7
Natural gas	45.5	100	108.4		57.2		36.6	
Coal	56.4	100	19.6		11.4		7.3	
Bioenergy and waste	0.7	100	2.3	328.6	5.9	842.9	7.9	1128.6

Table 2 UK production of primary fuels 1990 to 2014 (million tonnes of oil equivalent)

Numerical skills

Proportion, magnitude and ratio

Study Table 2, showing data on how the UK's production of primary fuels has changed in recent decades.

1. Complete the table by calculating the index figures for natural gas and coal. The base year is 1990 and so this is given an index of 100. For example, to calculate the natural gas index for 2000:
108.4 × 100 ÷ 45.5.

2. Use the index figures for oil, natural gas and coal to create a suitable graph to show the changes.

3. Bioenergy and waste as an energy source has increased by a huge amount, making it difficult to plot on a graph with the other primary fuels. Use single log graph paper to create a line graph to show all four primary fuels. Use the log axis for the production index.

4. Using your graph (or graphs) and the data in Table 2, describe what has happened to the UK's primary fuel production since 1990.

STRETCH IT!

UK energy facts

The UK's last deep coal mine closed in 2015. In 2014, 86 per cent of the UK's coal supplies were imported, mostly from Russia; 45 per cent of the UK's natural gas was imported, mostly from Norway; and 42 per cent of the UK's oil was imported from Norway and member countries of the Organization of the Petroleum Exporting Countries (OPEC).

Also in 2014, approximately 14 per cent of the UK's energy was from low carbon sources, 50 per cent from nuclear (but declining) and 30 per cent from bioenergy (increasing). The energy sector still accounted for most of the UK's greenhouse gas emissions, but total emissions are falling – down 36 per cent in 2014 compared with 1990 and down 25 per cent since 2005, showing a faster rate of decline over time.

DO IT!

North Sea oil and gas

Create five multiple-choice questions on how the UK's energy supply has changed. Then test a friend.

NAILIT!

Advantages and disadvantages

Make sure that you are able to give advantages and disadvantages of using the UK's main energy sources such as nuclear and wind energy, as well as the possible widespread use of fracking.

Figure 6 North Sea oil and gas fields

Economic and environmental issues

1   Non-renewable fossil fuels (oil, coal, natural gas) produce carbon dioxide when burnt; this greenhouse gas contributes to climate change. As supplies dwindle, the costs for consumers will increase. There are high costs involved with exploiting the more difficult to get at reserves and the environmental damage from fracking and opencast mining is extensive.

2   Nuclear power is controversial; while some people see it as a non-polluting source of energy, others are worried about accidents and the high costs involved in disposing of radioactive waste and decommissioning the power stations at the end of their productive life.

3   Wind energy is a renewable energy resource with low long-term financial costs and no carbon dioxide emissions once operating. Yet, wind farms are considered to be unsightly by some people due to their large size, and their construction can disrupt natural environments such as the seabed. However, there are jobs linked to **research and development (R&D)** of this future resource.

 # CHECKIT!

1   Explain one way in which people use water to improve their lives.

2   Explain why the patterns of food consumption vary around the world.

3   Explain why energy supplies are unevenly distributed around the world.

4   Explain two ways in which the demand for food in the UK has changed.

5   Describe how farming activities in the UK may contaminate freshwater supplies.

6   Explain the economic issues arising from energy use in the UK.

Food

Insecure food supplies

Areas of surplus and deficit

Food supplies (and prices) are affected by:

- **environmental factors**: climate change, ecosystem degradation, pests, disease and water scarcity

- **human factors**: population growth, income growth, changing diets, land scarcity, deforestation, political instability, technology and investment, and price shocks.

The areas of the world that do not have these risks and constraints will have a surplus of food (food security), while those with them will face deficits in food supplies (food insecurity).

Global patterns of calorie intake and food supply

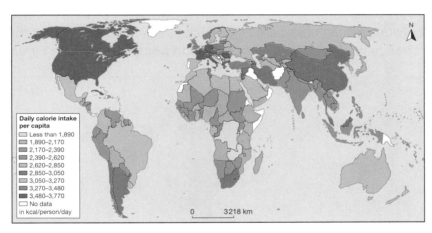

Daily calorie intake per capita
- Less than 1,890
- 1,890–2,170
- 2,170–2,390
- 2,390–2,620
- 2,620–2,850
- 2,850–3,050
- 3,050–3,270
- 3,270–3,480
- 3,480–3,770
- No data
in kcal/person/day

0   3218 km

Figure 7 World food consumption, 2015

Reasons for increasing food consumption

1. **Population growth**: more people means that more food will be required in the future. However, the world population rate is slowing gradually.

2. **Increasing wealth**: when people have more money, they buy more food and different types of food. More countries are developing economically with people becoming wealthier.

3. **Changing diets**: people's changing food tastes puts pressure on farmers and markets to supply the foods that are in demand, so fewer staple crops and more export crops may be grown.

Factors affecting food supply

Food security and food insecurity depend on:

- availability (own production or from markets)
- accessibility (ability to buy)
- safe use (clean water for food preparation)
- stability (long-term supply).

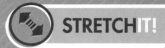

STRETCH IT!

Food balance

Latest information from the FAO suggests that there is a better balance of food supply in recent years, with overproduction of wheat and rice, with meat consumption stable and with fish production matching consumption.

DO IT!

Global food pattern

Look carefully at Figure 7, showing calorie intake per person per day. What patterns can you see? How many correct statements can you make about the patterns you see?

Food terminology

Make sure that you are able to use the correct food terminology:

- **Food security** means that there are enough affordable supplies of food, perhaps leading to a food surplus (more food than needed).

- **Food insecurity** means that there are not enough affordable supplies of food (a food deficit).

Also identify the reasons why a place has food security or insecurity – such as climate change making the area drier so that there is not enough water for crops.

Climate

Farming is largely weather dependent; there must be enough rain, warmth and sunlight at the correct times for food crops or fodder crops to grow well. If there is a drought, then there will be a shortage of food.

Technology

Production of food can be increased through the use of machinery and other inputs to the farming process, such as fertiliser. Research and development into new varieties of crop that are drought or pest resistant or better storage methods can increase the food available.

Pests and disease

All farm crops and animals may become diseased or be affected by a pest. Bananas have been affected by Panama disease, cattle in the UK have been affected by BSE and locusts still occasionally consume significant proportions of crops. Pests and disease may spread as climate change warms the world further.

Water stress

Farming is the biggest user of water in the world. Irrigation is important in the growing cycle of plants and, if there is a dry spell, then farmers take water to their fields. However, the demand for water for other uses, such as human consumption in urban areas, creates stress in some parts of the world. Climate change is also causing some areas to become drier (see Figure 8).

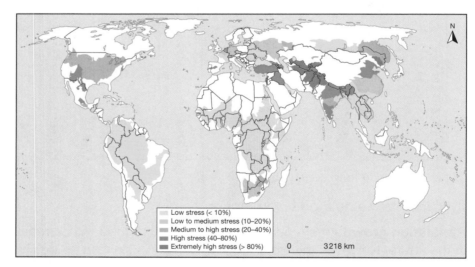

Figure 8 Water stress in the most populated world river basins

Conflict

Fighting between rival groups within countries, or between countries, disrupts the work of farmers so that they cannot plant or tend their crops, or look after their animals. In this way, whole years' worth of food supply may be lost.

Poverty

In LICs and NEEs there are still significant numbers of people who live on less than US$1.90 per day. If these people are landless, they cannot grow their own food and so must buy it. Often, they cannot afford food, especially at times of high prices.

The impact of food insecurity

Famine

A famine occurs when there is not enough food for the people living in an area. As a consequence, people starve. LICs in semi-arid parts of the world are more likely to experience this problem with high death tolls, for example, Somalia in 2010–12 and 2017.

Undernutrition

Undernutrition occurs when people are unable to obtain a balanced diet that provides all the vitamins and minerals that a healthy body requires. This may occur because people are too poor to buy a variety of foods or perhaps only staple crops like rice are available. Malnutrition weakens the human body so that diseases can kill and development in children is slowed.

Soil erosion

In an attempt to yield more food, farmers may use more chemicals or expand the growing area. Overuse of the land for farming or deforestation can expose soils to wind and rain erosion. The topsoil is the most fertile. If it is removed, crops will not grow well and there will be reduced crop yields. People may receive fewer calories per day than the human body requires.

Rising prices

If there is less food, then prices increase. The poorest people will be hit by this situation first and will have to go without food, leading to possible undernourishment and starvation.

Social unrest

Food is an essential for life, so when life is threatened, people get very worried and angry. This can lead to unrest among the people of a place, which can develop into protests, riots and even armed conflict, as people look for someone to blame or try to access food supplies. Conflict over other matters will disrupt farming, cause shortages of food and push up food prices, causing a humanitarian crisis.

Strategies to increase food supplies

Overview of strategies

Irrigation

Soil conservation and irrigation methods can tackle flooding or desertification and control soil erosion with the introduction of sustainable irrigation methods. Drip irrigation from small pipes is more efficient and conserves water better than open canals and ditches.

Aeroponics and hydroponics

Growing plants in misted air – aeroponics – or in water rich in nutrients – hydroponics – may be a solution, especially where there are land shortages as soil is not used (see Figure 9). However, these need a reliable water supply and enough money to construct the buildings and systems required.

Food security

Make a food picture diagram and divide it into two halves with a line. In one half of the diagram, list the things that may affect you getting food. In the other half, list the things that may affect someone living in an LIC getting their food.

Note the differences and think about why these differences exist.

Impact of food insecurity

Make a mnemonic out of the first letters of the impacts of food insecurity listed on this page. Use this to help you remember the impacts.

 STRETCHIT!

Recent staple food prices

Between 2004 and 2015, world maize and wheat prices increased by over 50 per cent and rice prices nearly doubled. Think about how this affects access to food.

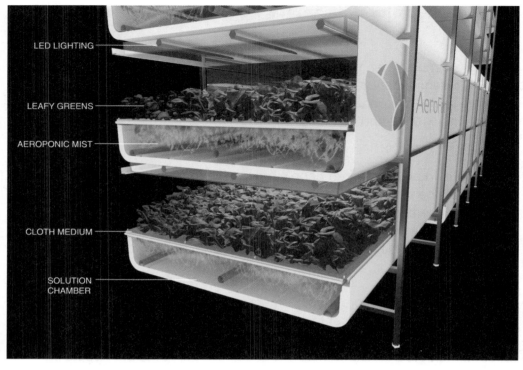

Figure 9 Aeroponics farming

New green revolution and biotechnology

- The recent green revolution uses resilient cultivation methods, such as water harvesting, soil conservation and irrigation combined with drought-resistant and salt-resistant crops, diversification into new crops and the development of seed exchanges. The aim is to increase yields (high-yielding varieties or HYVs) and also develop crops with a higher nutritional value.

- Genetic modifications are used to improve crops and animals, such as making crops pest, salt or drought resistant by introducing new genes from another plant or creature. Changing genes is controversial, as it may change the natural environment such as **food chains**, and some people are concerned about human health.

Appropriate technology

Appropriate technology means using natural methods as much as possible, with technologies that are a step up from basic tools, so avoiding expensive machinery. This would include drip irrigation rather than open channels, solar-powered water pumps or human-powered machines.

DOIT!

Think about the four strategies to increase food production. Decide on an order from best to least best strategy. Make sure that you can write down one reason for putting the best strategy first and one reason for placing the least best last.

STRETCHIT!

GM crops

Genetically modified (GM) crops have been controversial ever since they were first developed. Some people are concerned about how they break links in food chains and how the human body may be affected by eating foods with altered DNA. However, some countries, especially NEEs, see genetic modification as a definite way of increasing food supply. In 2015 the USA produced 71 million hectares of GM crops, Brazil 44 million ha, Argentina 25 million ha, India 12 million ha and Canada 11 million ha.

Case study

Example of a large-scale agricultural development: the Indus Basin Irrigation System (IBIS)

The River Indus is the main river flowing through Pakistan in southern Asia. The upper course flows through India. Since the Indus Water Treaty (1960), the river system has been developed into one of the largest irrigation schemes in the world. There are large and small dams, reservoirs, such as Tarbela (see Figure 10) and canals to transfer water to 14 million ha of irrigated farmland.

Advantages of IBIS

- Increased food production from a large irrigated area.
- Improved nutrition for people in Pakistan and India from the wider range of crops, such as fruits, that can be grown.
- Fish farming in the reservoirs, providing a protein source for people's diets.
- Higher yields have enabled surpluses to support food-processing industries and enable income from exports to be earned.
- Hydro-electric power from the largest dams has provided electricity for farmers' homes and local industries.

Disadvantages of IBIS

- Too much water can be extracted, meaning that farmers in the lower part of the river basin do not have enough.
- There is a large evaporation loss from the open reservoirs and large canals, reducing the amount available for farming.
- Poor irrigation techniques have led to salinisation of soils in places. This is where salts are drawn to the surface as water evaporates from fields and these salts kill crops.
- Sometimes water levels are so high that fields become waterlogged and crops will not grow.

Figure 10 Tarbela Dam on the Indus River

DO IT!

Large-scale agricultural development

Make a revision card listing three key advantages and three key disadvantages of a large-scale agricultural development.

Sustainable food supplies

Organic farming

This type of farming only uses natural fertilisers (manure or compost). These add moisture as well as nutrients to a soil. Organic farming practises crop rotation, so that different nutrients are used and replaced each year. It relies on the natural food web for biological pest control. Yields can be lower, but soils are looked after so that they can be used by future generations.

Permaculture

Permaculture involves self-sufficient farming. Its philosophy follows certain principles such as accepting nature's help with yields (e.g. bees), accepting that the Earth is self-regulating and adapting to its changes, using renewable resources and natural building materials and valuing biodiversity.

Urban farming

Some cities have urban farms, but more common is the use of allotments where vegetables and other produce are grown intensively by a family on a small plot of land. Even the balconies of flats or green roofs can be used for farming.

Sustainable sources of fish and meat

- Using net sizes that only catch mature fish, leaving enough to repopulate fishing areas and reducing the by-catch (those fish caught accidentally that are not wanted) are sustainable fishing methods. Sometimes fish quotas are set, which means that the fishing industry can only catch a certain amount in a year. Aquaculture, also known as aquafarming, is fish farming in netted areas or ponds, where fish, for example salmon, are reared from eggs to full size.

- Meat production is often land and water intensive due to the growing of fodder crops to feed the cattle or other farm animals. Therefore, using open grasslands for grazing animals in an extensive farming method is more sustainable, for example free-range chickens.

Seasonal food consumption

- Nature produces foods at different times of the year and these seasonal rhythms can be used as harvest times by people, e.g. fruits, summer berries, wheat/rice and migrations of fish and animals.

- Storage is an important part of this process to provide foods for less plentiful times of the year. In tropical climates, it may also be possible to obtain two or more harvests of staple crops such as rice each year.

Reduce waste and loss

- In high income countries (HICs), a lot of food is never eaten, either because it is past the use-by date or because consumers buy and cook too much. Changing consumer habits and reducing stocks would help to reduce waste, along with freezing perishable seasonal foods for later use.

- In LICs and NEEs, poor storage methods cause the loss of a lot of food. Better storage methods (dry and cool) would help to increase food security. Improved transport systems would get food to markets more quickly.

NAILIT!

Sustainability

Remember that **sustainability** is an idea that can be applied to many aspects of geography. It means making sure that resources like food are available for future generations.

NAILIT!

Advantages and disadvantages

Each sustainable method of farming has advantages and disadvantages. A common advantage is that they are all more environmentally friendly than normal farming, and a common disadvantage is that yields are likely to be lower.

DOIT!

Think about which is the best way of ensuring that there is a food supply in the future. Write down two reasons for the one that you think is the best.

Case study

Example of a sustainable local scheme: Jamalpur, Bangladesh

Sustainable food supplies

The farming of local **subsistence farmers** has been improved with the help of the **non-governmental organisation (NGO)** Practical Action. Before food supplies were made sustainable, the situation for families was very bad, with food supplies running out after about eight months every year and farmers having to borrow money to get food for the other four months. Hunger and malnutrition were common and families did not have enough money to send children to school.

- The variety of rice was changed in the paddy field system to one more resistant to flooding, so that yields were higher and more reliable.
- Small local fish were introduced to the paddy fields to provide a protein food source and to improve the paddy field system through adding natural fertiliser.
- Fruit trees and vegetables were planted on the dykes to provide a balanced diet.
- Most years, there is now a surplus of food, which can be sold at local markets. Some of this money is used to send the children to school and to make basic improvements to sanitation in homes.

Figure 11 Sustainable farming in Bangladesh

DOIT!

Local sustainable food scheme

Make a revision card listing four key points about a sustainable local food scheme. Make sure that you concentrate on the food aspects.

CHECK IT!

1 Give a correct definition of the term 'food insecurity'.

2 a Describe three reasons for the increase in world food consumption.

  b Explain how crop pests and disease may affect food supply.

3 Explain three impacts on people in a country experiencing food insecurity.

4 Describe the main features of aeroponics.

5 a Explain the aims of permaculture.

  b Explain why using biotechnology to increase food supplies is controversial.

6 a Describe two advantages of a large-scale agricultural development that you have studied.

  b Choose one feature of a sustainable small-scale food supply scheme that you have studied. Explain how this feature has helped to make the scheme sustainable.

Water

Insecure water supplies

Areas of surplus and deficit

Water stress exists in parts of the world where the majority of the water available is used up every year (see Figure 12). This situation exists due to natural supplies being too low and human consumption too high. Countries with the highest level of water stress include Libya, Pakistan and Saudi Arabia.

Global patterns of water surplus and deficit

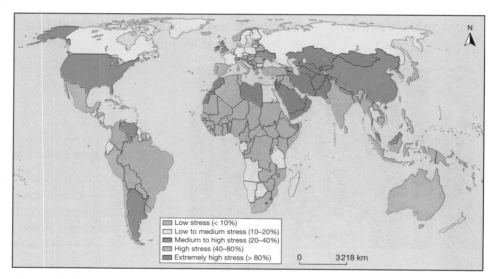

Figure 12 Water stress by country

DO IT!

Global pattern of water supply

Look at Figure 12, showing areas of the world that have lots of water and those that have little.

Write a list of those areas that have a lot of water and another list of those that have little. (Use names of continents, world regions, countries, or parts of countries.)

Do the areas in each list have anything in common that influences the amount of water they have?

Reasons for increasing water consumption

- **Increasing population**: the world population continues to grow and each person needs a quantity of clean water to drink in order to live, but also to support the other activities in their lifestyle. People also need more food, and farming is the biggest user of water.

- **Economic development:** more industries and businesses mean that more water will be needed for use in processing products, cooling machinery or irrigating commercial agriculture.
- **Better standard of living:** as people become wealthier, they use more water in their daily lives, perhaps because they understand the need for hygiene. As more people live in urban environments, the demand for water increases.

Factors affecting water supply

Climate

There are different climates around the world and, clearly, there are greater supplies of water in wetter areas than drier areas. A major problem is climate change, which is altering precipitation patterns and making some areas wetter and some areas drier.

Geology

The amount of water on the surface and in the ground depends partly on the rocks. If the rock is impermeable, more water will flow over the surface carrying water to other areas and also allowing reservoirs to be created. However, water will not soak into the ground so groundwater supplies are low. Porous and permeable rocks allow water to pass into them so sometimes there is not much surface water but aquifers are likely to form underground.

Pollution

Everyone needs fresh clean water. Therefore, if water is polluted, it cannot be used. Unfortunately, where people are located, so too are industries and farming, and contamination of water is always possible. In more remote rural areas there may not be a proper sewerage system, and rivers and groundwater may become contaminated.

Over-abstraction

There will be a certain amount of water in rivers or in the ground. Too much water may be taken out by farming, industry or urban areas, so that there is not enough left to go around.

Limited infrastructure

In order for people to receive water, there needs to be a system of extraction, cleaning and transport from the water stores to the consumer. In HICs, these systems are well developed, but in LICs and NEEs the urban areas may only have a partial system (or none in squatter settlements). Rural areas may have none because it is too expensive for the country to provide (see Figure 13).

Figure 13 Water collection in an LIC

Water availability

Draw a table with two columns: one headed 'HIC' and the other 'LIC'. In each column, add positive and negative influences on water availability in that type of country.

STRETCHIT!

Water disputes

As well as disputes over the quantity of water in rivers, there may also be disputes over pollution moving downstream from one country to another. For example, there have been major incidents on the rivers Rhine and Danube in continental Europe.

DOIT!

Plan an answer to a question that asks about the impact of shortages of clean fresh water on an NEE.

Poverty

In developed countries, people often pay water rates or have a water meter so they can be charged for what they use, and most people can afford these costs. However, in LICs and NEEs, a lot of poorer people have to spend many hours each day collecting and carrying water from a single communal source or river. In rural areas this often involves walking a long way.

Impact of water insecurity

STRETCHIT!

Multiple water uses

In many LICs and NEEs, rivers have multiple uses – water supply, irrigation water, washing and bathing, disposal of sewage, disposal of industrial waste and toxic chemicals, disposal of farm waste, transport, hydro-electric power and may also have religious significance (for example, the Ganges). These uses clearly conflict, and the health of people is seriously affected. Safe drinking water is essential for people, but nearly 800 million people do not have safe water.

Disease and pollution

- Water in areas where sewage is not treated becomes contaminated with harmful bacteria, which cause diseases such as dysentery or cholera.
- Water may also become polluted by chemicals such as mercury, phosphates or oil, from industries or farming. These chemicals can damage human health directly or through the eating of contaminated fish.

Food production

Without water for irrigation, crop yields may decline, limiting people's access to food and possibly leading to malnutrition and starvation.

Industrial output

The production from factories, HEP stations and businesses may be affected by water shortages for processing and cooling machinery, which will, in turn, harm the economic progress of a region or country.

Potential conflict

Where supplies are limited, water becomes a valuable commodity, and is essential for life, so people may fight over it. This conflict may involve different countries where the water source is shared, such as a river flowing through several countries, especially when dams are built or proposed upstream of a country.

Strategies to increase water supplies

Overview of strategies

The main disadvantages of most strategies are the high financial costs of construction and the potential damage to the natural environment, such as flooding areas or changing the discharge of rivers.

Diverting supplies and increasing storage

Water supplies can be diverted from rivers, such as taking water from the Colorado River in the USA and diverting it through canals to southern California. Arizona uses surplus water from the Colorado River to recharge its aquifers underground, ready for times of water deficit.

Dams and reservoirs

Water can be stored along a length of a river by building dams and creating reservoirs, such as Lake Mead behind the Hoover Dam on the Colorado River in the USA.

Water transfers

Water can be moved between river basins, taking water to areas that have the greatest need, such as from the Yangtze in southern China to the Huang He in northern China (see case study below).

Desalination

The process of **desalination** involves taking seawater and separating the salts from the water to produce fresh water; a lot of energy is used in the process, so it is expensive.

Case study

Example of a large-scale water transfer scheme: China – south to north water transfer

Southern China is a humid region with enough water to meet the needs of the area, but north-eastern China is drier and has many large urban areas, farms and industries that desperately need more water. Groundwater supplies are running out in the north-east and something needs to be done. Two water transfer routes have been completed or partially completed: a transfer from the lower Yangtze River northwards through existing canals, rivers and lakes; and a transfer from the central part of the Yangtze River basin northwards to the Huang He River and the capital Beijing (see Figure 14).

Advantages

- Huge amount of water transferred to areas of water scarcity and stress (45 billion m³).
- Cities and industries of the north-east receive the water that they need.
- Supports the economic development of China.

Disadvantages

- Huge cost of completing the scheme (more than US$80 billion).
- Many Chinese waterways are polluted and the water transferred is likely to be contaminated.
- Many people displaced to make way for new reservoirs.

NAILIT!

Indus Basin Irrigation System (IBIS) scheme

Recall detailed information about the scheme in Pakistan and India (see page 157).

NAILIT!

UK water transfer

Recall detailed information about water transfers in the UK (see page 149).

STRETCHIT!

Desalination

The countries with the largest desalination capacity are Saudi Arabia, the USA and the UAE. Saudi Arabia produces 5 million m³ of fresh water a day through desalination. It is worth noting that these three countries are all oil rich!

- River flows have been changed.
- The Yangtze River's transportation system may be affected.
- Wildlife and ecosystems have been affected.
- Droughts in the south will reduce the amount of available water in some years.
- Considerable evaporation losses from open channels and canals.

Figure 14 China's water transfer scheme

Sustainable water supplies

Conservation

- Water can be saved by using it more efficiently; for example, making sure that time spent in the shower is short, using short cycles in a washing machine and installing water meters to make people aware of what they are using.
- Farmers can use the most efficient types of irrigation such as drip irrigation and industries can prevent pollution to conserve fresh water.
- Water companies can reduce leakages from pipes in their systems.

Groundwater management

- Many areas have rock layers underground that store water (aquifers). People, water companies and industries can drill down into the aquifers to create wells or boreholes from which to pump water. However, if too much water is taken out, levels drop and supplies are threatened.
- **Abstraction** can be managed by governments only giving permission for a certain amount of water to be used, enforcing pollution regulations, and by finding methods of replacing the water used (recharge), perhaps by transferring water from areas that have a surplus.

Recycling

- Rainwater or water that has been used for washing can be stored and then reused again for gardens, washing vehicles or flushing toilets.

- Once water has been through sewage works and harmful bacteria have been removed, it could be used for the irrigation of fields, fish farms, to recharge groundwater or for cooling machinery.

'Grey' water

Grey water is water that is dirty but does not contain harmful bacteria or chemicals, such as that created by washing or rainwater that has been collected. Grey water can be filtered and used for watering gardens or flushing toilets.

> **DO IT!**
>
> **Local sustainable water scheme**
>
> Create a revision card summarising four key points about an example of sustainable water supplies that you have studied. Make sure that you concentrate on the water aspects.

Case study

Example of a sustainable local scheme: Hitosa, Ethiopia

Sustainable water supplies

Hitosa is a rural area in central Ethiopia, located in a hot and dry area. Traditionally, water was collected from the few small seasonally flowing rivers and one spring. In the 1990s, a sustainable water scheme was developed to move water from permanent springs on the slopes of a local mountain, Mount Bada, by using gravity and 140 km of pipeline to 100 stand taps and 150 farmers. Over 65 000 people now have access to 25 litres of water per person a day.

The NGO WaterAid provided over half of the money for the scheme and provided expertise to help design and implement it. The scheme is now managed by local communities and people have to pay a small amount of money towards the maintenance costs of the scheme. This is a typical bottom-up project where decision making is made at a local level with the help of a charity and, therefore, is more appropriate to the long-term needs of the local people.

The scheme could have been improved with health and hygiene education and limitations on the amount of water used by farmers. Future challenges include population growth and, therefore, pressure on the water supplies and the replacement of the pipeline when it gets old and deteriorates.

✓ CHECKIT!

1 Give a correct definition of the term 'water insecurity'.

2 a Describe three reasons for the increase in world water consumption.

　b Explain how over-abstraction may affect water supply.

3 Explain three impacts on people in a country experiencing water insecurity.

4 Describe the main features of desalination.

5 a Explain the aims of increasing water storage underground.

　b Explain why using dams and reservoirs to increase water supplies is controversial.

6 a Describe two disadvantages of a large-scale water transfer scheme that you have studied.

　b Choose one feature of a sustainable local water supply scheme that you have studied.

　c Explain how this feature has helped to make the scheme sustainable.

Energy

Insecure energy supplies

Areas of surplus and deficit

An **energy gap** exists in parts of the world where energy supplies are not enough to meet energy demands. This situation exists due to energy supplies being too low or human consumption too high. The country with the largest energy gap is India.

DO IT!

Global pattern of energy insecurity

Look at Figure 15, showing which areas of the world have high and low levels of energy access and security.

Write a list of areas that have enough energy supplies and another list of those that do not have enough energy. (Use names of continents, world regions, countries, or parts of countries.)

Do the areas in each list have anything in common that influences the amount of energy they have?

NAIL IT!

Energy terminology

Make sure that you know and can use key terms associated with energy:

- **energy consumption** is the amount of energy consumed by a country or area, usually given as an amount per person
- **energy supply** is the amount of energy produced by a country or area, usually given as an amount per person
- **energy security** exists where a country or area produces enough energy to meet its needs
- **energy insecurity** exists where a country or area does not have enough energy to meet its needs
- **energy gap** is the difference between the amount of energy available and the amount of energy needed or consumed.

Global patterns of energy supply and consumption

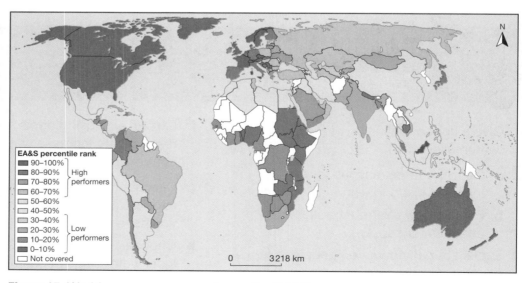

Figure 15 World energy access and security (EA&S)

Reasons for increasing energy consumption

Global energy consumption tripled between 1965 and 2015 and is expected to double again by 2040. The reasons for this are:

- **Economic development**: with developed countries being caught up by the NEEs, there is a lot more industry and business in the world, an intensification of farming and greater use of transport. Globalisation has also expanded trading and economic links. All of these activities use energy in greater quantities.

- **Increasing population**: the more people there are, the more energy gets used, especially as people become wealthier and can afford to use more electricity and transport. They also buy more consumer products, which means more industrial activity takes place (see Figure 16).

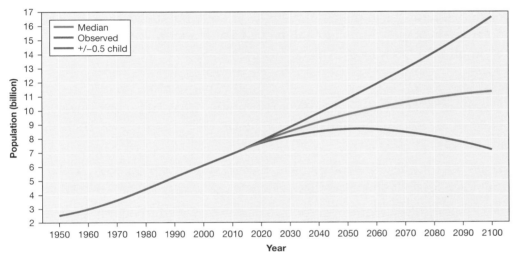

Figure 16 World population growth and predictions

- **Technology use**: the availability and range of technology in people's homes and within countries has greatly increased since 1965. These technologies often use electricity and require national electricity grids and fuels for the power stations. Car ownership has also rapidly increased, so the demand for oil has greatly increased.

Factors affecting energy supply

Physical factors

- **Geology**: fossil fuels and uranium were formed by geological processes and are only found in certain places, such as the large oil reserves in the Middle East, or geothermal energy near plate boundaries, such as in Iceland.

- **Climate**: some renewables depend on weather and climate factors, such as enough wind for wind farms and enough sunlight for solar farms. Countries with a coastline with land-shore breezes, exposed mountains or in the mid-latitudes (halfway between the equator and pole) where there are low pressure systems (depressions) have wind energy potential. Countries with clear skies have the potential for solar energy, ranging from deserts to countries with a calmer atmosphere such as Germany.

- **Environmental factors**: a combination of severe climate and difficult access makes some energy reserves difficult to get at, such as oil fields under the Arctic Ocean or in Siberia.

Reasons for increasing energy use

Draw a flow diagram to show how changes to the economy, population and technology have led to an increase in energy use.

Costs of exploitation and production

- The value of an energy source is linked to demand and the amount of it; if it is in demand or is running out, the value goes up and then it is worth companies looking for more of that resource. This has happened with oil as there is a strong dependence on it at the moment.

- The cost of converting a primary energy source to electricity is another consideration, as are the high costs of refining oil into petrol and diesel.

Technology

- Changes in technology change the type of resource required. For example, steam engines needed coal, while motor vehicles need oil. There are many electrical appliances and gadgets today that require more electricity.

- Extraction technologies have also changed, with an ability to drill for oil and natural gas in deeper oceans or to carry out fracking. Wind and solar technologies continue to improve to offer renewables as a genuine alternative to fossil fuels in electricity production.

Political factors

- Countries will try to be self-sufficient in energy production and develop policies to support this aim, such as the UK giving permits for fracking and investing in wind farms.

- Countries may also depend on being friendly with other countries, so that they can trade for their energy resource and use political alliances to secure them.

- In LICs and NEEs, corruption or civil unrest may disrupt energy supplies.

- Environmental laws and aims are being set internationally, which requires countries to meet sustainable and environmental targets in energy use and production.

STRETCH IT!

Government energy policies

The governments of Germany and Japan have changed their minds about using nuclear energy after the disaster at Fukushima in 2011 because of the realisation that radioactivity is potentially very dangerous. The UK government had promoted solar and wind energy with subsidies, but then withdrew them, stating that these industries needed to become competitive in the energy market and that renewable energy costs were being passed on to consumers. The 2008 **recession** also put pressure on government budgets.

DO IT!

Factors affecting energy supply

Summarise the factors affecting the availability of energy supplies in 50 words. Think about whether the physical or human factors are more important.

Net importer of energy	Million tonnes oil equivalent	Net exporter of energy	Million tonnes oil equivalent
China	508	Russia	571
Japan	422	Saudi Arabia	406
India	290	Australia	235
USA	258	Indonesia	232
South Korea	233	Canada	185
Germany	197	Qatar	174
Italy	115	Norway	167
France	114	Kuwait	131

Table 3 Highest net importers and net exporters of energy, 2014

Graphical skills

Imports and exports of energy

Using the data in Table 3 and a blank outline map of the world, draw bars over the relevant countries on the map using a scale (perhaps 1 cm = 100 million tonnes oil equivalent) to show net energy imports (yellow) and net energy exports (green). Make sure that you give a title, key and scale for the bars.

Does the pattern match Figure 15?

Impact of energy insecurity

Exploration of difficult and environmentally sensitive areas

If a country is short of energy (or money for development), it may be encouraged to extract resources from any possible source. This may lead to:

- The clearance of natural ecosystems such as boreal forests in Canada or tropical rainforest in Brazil, or oil and gas exploration in cold environments such as the Alaskan tundra, the Arctic Ocean or deep water in the Gulf of Mexico.

- Hydro-electric power schemes requiring dams and reservoirs to be constructed in hilly or mountainous areas; these flood large areas of an environment.

- Renewable schemes such as wind or solar farms being placed in areas of high scenic value, 'spoiling' the scenery.

Figure 17 Canadian tar sands exploitation

Economic and environmental costs

All types of business will be affected if there is not enough energy. The economy of a country will decline and power cuts may become a regular occurrence – in Brazil, for example. This may lead to:

- people facing higher energy bills and being without electricity in their homes at times and industries not being able to operate at full capacity

- people in LICs and NEEs cutting down more trees for firewood, causing deforestation and soil erosion

- a reduction in pollution, but it may encourage risky exploration and developments such as fracking (oil and gas) or drilling for oil in deeper water, which increases the risk of oil spills.

Food production

Farming may not be intensive enough to produce food for a country if there is not enough fuel for machinery or for factories producing agricultural chemicals. This may lead to:

- higher food prices for people due to shortages or expensive imports

- a pressure to grow biofuels, which reduces the area available for food crops

- people in LICs having to spend a lot of time searching for firewood, meaning that they have less time to work on their fields.

Industrial output

Industries may have to reduce production if there are power cuts or fuel shortages. This may lead to:

- businesses and industries making less money and causing unemployment

- price increases or an increased dependence on imported products

- changes in the trade balance, with more imports than exports.

Potential conflict

If energy supplies are insecure, there may be conflict within a country or internationally, such as:

- between consumers if there is not enough energy for all, with higher prices

- between businesses, industries and the government within a country with high energy costs

- between economically competing countries if one is keeping energy costs lower than they should be, giving their industries an unfair advantage

- over areas with key energy sources, such as the Middle East; new energy reserves, such as the Arctic Ocean; or important routes for the transfer of energy, such as the Black Sea area.

DO IT!

Create a revision diagram showing how energy insecurity can affect many geographical aspects of a country.

Strategies to increase energy supplies

Overview of strategies

Renewables

Resource type	Positives	Negatives
Biomass	• Renewable if new vegetation is continually planted • Affordable for poorer people • Can be used in large and small power stations	• Releases carbon dioxide into atmosphere • Can lead to deforestation • Reduces land area for food crops
Geothermal	• Cheap energy source • No carbon dioxide emissions • Can generate electricity, heat homes and provide hot water	• Limited to certain locations where magma is close to the surface • Needs expensive and complex technology • Only the electricity produced can be transported long distances
Hydro	• Produces very cheap electricity • Flexible power generation to meet demand • Creates a reservoir for water supply	• Very expensive to build • Can destroy a large natural area • Can displace many people from their homes
Solar	• Low maintenance required once installed • No noise pollution • No carbon dioxide emissions during electricity production	• Very large solar farms may be required, using up farmland • Metals and materials used in manufacture can harm environment • Some consider them a visual pollutant
Tidal	• Countries with a coastline of estuaries and bays can use this source • No carbon dioxide emissions • Electricity generated twice every day	• Only a few locations have a large enough tidal range • The tidal barrage interferes with nature • The construction may be large and unsightly
Wave	• Countries with coastline and long fetches can use this technology • Can use the motion of small waves • No carbon dioxide emissions	• Technology not yet proven to be reliable • Power supplies dependent on wave conditions which vary a lot • Cannot be used by landlocked countries
Wind	• Free energy source, so cheap electricity • Multiple wind turbines create flexibility to generate different amounts of energy • Can use inexpensive micro-turbines for individual farms or homes	• Wind turbines can be very large and obtrusive • Extra costs involved if located offshore • Reportedly endangers wildlife

Table 4 Renewable sources of energy

Non-renewables

• **Fossil fuels:** conventional power stations can be made more efficient by reusing the heat created during the generation of electricity. This combined heat and power system reduces the amount of fossil fuel used and can produce up to 50 per cent more energy. Conventional power stations can burn biomass as well as coal and natural gas, making electricity generation cheaper and slightly more eco-friendly. Using natural gas rather

Renewable energy sources

Put the seven renewable sources of energy listed in Table 4 into rank order according to how successful they would be at increasing future energy supplies.

Give three reasons for your top and bottom choices.

than coal or oil is better as it produces lower quantities of greenhouse gases. Capturing carbon dioxide emissions and storing them underground makes power stations more environmentally friendly for the future.

- **Nuclear:** used nuclear fuel (uranium rods) can be reprocessed so that uranium can be retrieved to use again. But some HICs have decided that the potential danger from radiation is too much of a risk to take.

Case study

Example of fossil fuel extraction: Amazonia natural gas

In Peru, the Camisea project was started in 2004 to extract natural gas reserves estimated to be 385 billion m³. These reserves are expected to last until 2034. The project was financed by foreign sources and the operations are also mostly foreign owned. Pipelines take the natural gas to a processing plant near Cuzco and, from there, pipelines take the gas to Lima, the capital, and to the port of Pisco where the gas is prepared for export – mostly to Mexico.

Exports began in 2008. A pipeline to take the gas to southern cities and copper mining areas of Peru is under construction. The natural gas reserves will supply 95 per cent of Peru's needs. Further exploration of the area for more reserves continues. However, the area has been described as a 'fragile biodiversity hotspot' and is also the home of indigenous tribes, such as the Nanti. In 2014, lawsuits were filed to stop expansion into the national park and indigenous areas of the Kugapakori-Nahua-Nanti Reserve and, in 2016, presidential election debates often focused on the benefits and problems of the Camisea gas resource.

Advantages

- Peru has a cheap source of fuel with consumers saving US$13.7 billion in electricity costs between 2004 and 2014.
- The companies producing, transporting and selling the gas have paid about US$8 billion to the national and regional governments in taxes and royalties since extraction started.
- The gas reserves are contributing 0.08 per cent to Peru's gross domestic product (GDP) growth each year, and are predicted to add between US$23 billion and US$34 billion to Peru's economy before the reserves run out.
- Energy security has been obtained for the country, with natural gas supplies and gas-powered electricity-generating stations offering controllable production.

Disadvantages

- Tropical rainforest has been cleared for the gas operations and pipeline construction (see Figure 18), including areas set aside as parks and reserves.
- Indigenous tribes have been disrupted, losing their home areas and being infected with illnesses from the outsiders.
- Camisea natural gas is a finite resource and will run out.
- The exploration, production and distribution of the natural gas is very expensive, beyond the means of the country, so money had to be borrowed and foreign companies brought in.

Figure 18 Damage to the tropical rainforest by natural gas pipeline construction in Peru

Photographic interpretation skills

Gas pipeline construction

Make a labelled sketch of Figure 18 to show all the changes caused to the natural environment by the pipeline construction.

Sustainable energy use

Sustainable energy use means reducing pollution, using renewable options and improving energy efficiency. This includes reducing individual energy use and carbon footprint.

Conservation

Conservation involves reducing our use of energy, for example, by walking or cycling and making sure that the energy that is used is used efficiently, for example, a condensing boiler.

Home design

All buildings, especially people's homes, can be designed to conserve energy by:

- insulating roofs and walls
- double- or triple-glazing windows with thermal glass
- using energy efficient appliances (A+++ rating)
- fitting solar panels, micro-wind turbines and ground-source heat pumps
- changing lifestyle habits to include switching off appliances when not in use, using electronic gadgets less frequently and turning down central heating thermostats (and wearing more clothing instead).

Workplace practices

All business premises can do similar things to 'home design', in addition to the following:

- Lighting in common areas to switch off automatically when not in use.
- Workers to be encouraged to cycle or car share to work.
- Unused spaces, such as storerooms and corridors, to be left unheated.
- Natural lighting via windows to be used rather than electric lighting.
- Heating and air conditioning to be used to regulate temperatures to legal levels only.

Sustainable transport

Walking and cycling are suitable for shorter journeys as no energy is used. However, for some people, and for longer journeys, this may not be possible, so alternatives include:

- using public transport
- using small energy-efficient cars
- car sharing when making the same journey
- using cars that run on ethanol, a biofuel.

Fossil fuels

Remember that fossil fuels are those formed a long time ago by geological processes, where the remains of marine plants and animals were compressed and heated underneath ancient oceans and seas to become part of the rock structure. These fossil fuels are coal, oil and natural gas, and are only found in certain locations where the geological processes took place.

Example of fossil fuel extraction

Produce a revision card summarising the advantages and disadvantages of an example of a location where fossil fuel extraction is taking place.

DO IT!

Sustainable energy use

Decide what you think is the best way of making energy use more sustainable in the future. Write down at least two reasons to back up your argument for the one that you think is the best.

Reduce demand

With increased wealth and availability of personal transport and technology, energy use has increased. In order to reduce demand:

- people need to change their attitude towards energy use, making an effort to use less (see Figure 19)
- governments need to pass laws to make manufacturers use energy-efficient technologies during manufacturing
- governments can charge different amounts of road tax depending on the eco-friendliness of the vehicle.

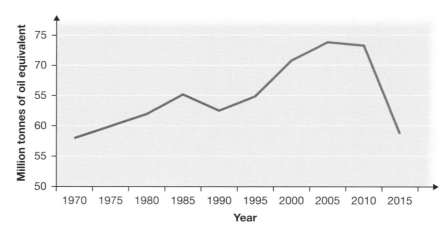

Figure 19   Domestic energy use in the UK, 1970–2015

Efficient technology

- Cars have become more efficient, with improvements in engine technology and weight reductions. Hybrid engines have reduced the amount of petrol or diesel burnt during use, and electric cars may be a further step forward (depending on how the electricity they use is created).
- Combined heat and power can make power stations more efficient by over 30 per cent.
- Household appliances are now more efficient at using energy.

DO IT!

Local sustainable energy scheme

Create a revision card summarising four key points about an example of a local sustainable energy supply that you have studied. Make sure that you concentrate on the energy aspects.

NAIL IT!

Large-scale or small-scale schemes

Most schemes involving resources can be divided into two groups: top down and bottom up. Top-down schemes are large, expensive and use a high level of technology, and are decided by a national government and international corporations or governments, such as the natural gas exploitation at Camisea in Peru. Bottom-up schemes are usually small, less expensive, use intermediate technology, and are decided by local communities with the help of NGOs, such as the micro-hydro at Chambamontera in Peru.

Case study

Example of a local renewable energy scheme: Chambamontera, Peru

Sustainable energy supplies

In the Andes Mountains of Peru there are steep slopes, poor road networks and isolated farming communities. It is therefore too costly to provide grid electricity, and even if it were available, the poor people would not be able to afford it. Instead, they use fuelwood for cooking and heating and kerosene lamps for light. There is plentiful rain with rivers flowing in the valleys, therefore a sustainable energy option is micro-hydro-electric power. Micro-hydro is suitable for small rivers, does not inflict significant damage on the natural environment and is a renewable source of energy. It works by diverting water at an intake weir through constructed channels on the mountainside to a storage tank; the water then runs through a pipe down a slope through a turbine which is turned to generate the electricity. The water then returns to the river. As there is no dam, there is no flooding and costs are very low.

The poor people still need financial and planning help, and this has been provided by the NGO Practical Action. One community to benefit has been Chambamontera, which has about 60 families. A micro-hydro scheme was started in 2008, costing US$45 000 and generating 15 kW of electricity. Most of the funding was from international aid but 6 per cent came from the village families. Chambamontera micro-hydro is an example of a 'bottom-up' scheme where the local community decided what they needed and now run the scheme.

Benefits

- Cheap renewable electricity is created (US$0.14 per kW).
- Energy is provided for homes and small businesses – including household businesses.
- There is little damage to the environment as there is no reservoir.
- Deforestation is reduced as fuelwood is not needed and side effects such as soil erosion are reduced.
- Kerosene (a fossil fuel) use is reduced and so the health of people in their homes has improved.
- Energy is provided for the school (computers and internet) and health centre (refrigeration).

Problems

- Poor people still have to pay for the electricity and any initial loans that they had arranged.
- The scheme has a 25-year lifespan, so eventually will need new investment (equipment and maintenance).
- There is some visual pollution on the sides of the valley and some alteration to the flow of the river.

CHECKIT!

1 Give a correct definition of the term 'energy insecurity'.

2 **a** Describe three reasons for the increase in world energy consumption.

   b Explain how the costs of exploitation and production may affect energy supplies.

3 Explain three impacts on people in a country experiencing energy insecurity.

4 Describe the main features of biomass energy.

5 **a** Explain the positives and negatives of wind energy.

   b Explain why using HEP to increase energy supplies can be controversial.

REVIEW IT!

The challenge of resource management

Resource management

1 Give a correct definition of the term 'malnutrition'.

2 Explain how energy resources help to improve the lives of people.

3 a Describe how soil fertility affects food supply in an area.

   b Explain how rapid population growth may affect the amount of food supply for a country.

4 Explain why food consumption usually increases as a country develops economically.

5 a Describe how mountain ranges affect water supply in an area.

   b Explain why high population densities put pressure on local water supplies.

6 Explain how climate change may reduce water supplies in some parts of the world.

7 a Describe how geothermal energy is created.

   b Explain the conditions necessary for solar energy.

8 Explain why world demand for energy has increased over time.

9 a Give a correct definition of the term 'agribusiness'.

   b Describe the role of supermarkets in the UK's changing food supply.

   c Explain how imported foods are considered to increase the UK's carbon footprint.

10 a Describe the role of leisure activities in increasing water demand in the UK.

   b Describe one way in which the UK government tries to ensure that people receive high water quality.

   c Explain why the UK transfers water from one part of the country to another.

11 a Give one reason why the UK's energy mix is changing.

   b Explain why our use of the UK's fossil fuel resources is in decline.

   c Explain why environmental concerns have arisen in the UK from plans to develop non-fossil fuel resources.

Food

1 a Name a country or world region that faces food insecurity.

   b Describe how climate affects the availability of food.

   c Explain why people in an LIC may have a problem getting access to food.

   d Explain how conflict in a country may affect the stability of food supply.

2 a Give a correct definition of the term 'famine'.

   b Explain why malnutrition is sometimes described as the world's biggest killer.

3 Explain why using appropriate technology in farming may be the best way of increasing food supplies in an LIC.

4 a Give two features of a large-scale agricultural development that you have studied.

   b Describe three advantages of this large-scale scheme.

   c Explain three disadvantages of this large-scale scheme.

5 Describe two features of organic farming.

6 a Explain how reducing food waste can make food supplies more sustainable.

   b Explain why using sustainable sources of fish and meat is important for future world food supply.

7 a Describe two features of a sustainable local agricultural scheme that you have studied.

b Explain how changes made by this local scheme have increased the sustainability of food supplies.

c Explain how these changes have improved the lives of local people.

Water

1 a Name a country or world region that faces water insecurity.

b Describe how climate affects the availability of water.

c Explain why people in an LIC may have a problem getting access to water.

d Explain how conflict in a country may affect the reliability of water supplies.

2 a Give a correct definition of the term 'water stress'.

b Explain how pollution affects water supply.

3 Explain why diverting water supplies from rivers may be the best way of increasing water supplies in an LIC.

4 a Give two features of a large-scale water transfer scheme that you have studied.

b Describe three advantages of this large-scale scheme.

c Explain three disadvantages of this large-scale scheme.

5 a Describe two features of water recycling.

b Explain how using grey water can make water supplies more sustainable.

c Explain why conserving water is important for future world water supply.

6 a Describe two features of a sustainable local water scheme that you have studied.

b Explain how changes made by this local scheme have increased the sustainability of water supplies.

c Explain how these changes have improved the lives of local people.

Energy

1 a Name a country or world region that faces energy insecurity.

b Describe how geology affects the availability of energy.

c Explain why people in an LIC may have a problem getting access to energy supplies.

d Explain how political factors may affect the security of energy supplies.

2 a Give a correct definition of the term 'energy exports'.

b Explain how climate affects energy supply.

3 a Explain why environmentally sensitive areas may be under increased threat if there is energy insecurity.

b Explain the positives and negatives of using renewable tidal energy in a country's energy mix.

4 a Give two features of a fossil fuel extraction scheme that you have studied.

b Describe three advantages of this fossil fuel scheme.

c Explain three disadvantages of this fossil fuel scheme.

5 a Describe two features of home design that reduce energy use.

b Explain how energy demand can be reduced to make energy supplies more sustainable.

c Explain why using sustainable transport is important to future world energy supplies.

6 a Describe two features of a local renewable energy scheme that you have studied.

b Explain how changes made by this local scheme have increased the sustainability of energy supplies

c Explain how these changes have improved the lives of local people.

Issue evaluation

THE EXAM!

- This section is tested in Paper 3 Section A.
- A pre-release resource booklet is issued 12 weeks before the exam.

Introduction

You will have studied and revised geographical topics separately for the first two exams, but in the real world everything is interlinked and the Issue evaluation question paper is a way to test your ability to see the wider picture. Complicated situations arise in real life, where the links between human and physical geography and the processes within them result in problems or difficulties that require decisions to be made to solve or reduce them.

- **Climate change**: as temperatures get warmer, there will be more tropical storms. Areas where it is wetter may become flooded and other areas could be drier with an increased risk of **drought**.

Figure 1 Resource management links

DO IT!

Energy issues

Look at Figure 1. Make a list of all the issues you can spot.

NAIL IT!

Problem solving

This is the process of finding solutions to geographical issues or problems.

STRETCH IT!

Making links

Study Figure 1, especially the links between the boxes. On a copy of Figure 1, add other boxes and arrows to show all other geographical links.

Figure 2 Cold environment management links

Making decisions

Study Figure 2. Think about the attitudes and opinions of stakeholders at the international, national and local levels. Suggest the difficulties that are created by contrasting attitudes and opinions in the decision-making processes that manage cold environment areas.

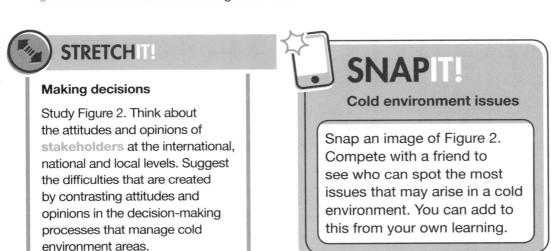

Cold environment issues

Snap an image of Figure 2. Compete with a friend to see who can spot the most issues that may arise in a cold environment. You can add to this from your own learning.

Attitudes and opinions

Look at Figure 2. Think about the variety of attitudes and opinions that may exist in the management of cold environments. Note down the reasons why people, governments or organisations (the stakeholders) have contrasting attitudes and opinions regarding the situations in cold environments.

How to revise for Issue evaluation

Revision and preparation for Issue evaluation will be a little different from Units 3.1 and 3.2. Papers 1 and 2 are based on the physical and human geography topics, but Section A of Paper 3 (Issue evaluation) could be on any topic or combination of topics.

In the Issue evaluation exam, you will be given a real issue to consider and this will be outlined in a resource booklet, which is made available to teachers 12 weeks before the Paper 3 exam. Your teacher will give you the resource booklet for your exam when they have prepared lessons for looking at it. Your main knowledge and understanding revision for Paper 3 will be completed when you revise the topics within 'Living with the physical environment' and 'Challenges in the human environment'.

What you need to concentrate on is:

- your ability to answer the style of questions that appear in Issue evaluation

- developing familiarity with the material contained within the resource booklet

- making sure that you understand the links between all of the geographical ideas contained within the resource booklet information.

Paper 3 revision schedule

Find out from your teacher when you will be given the resource booklet, how many lessons you will spend on it and what you will be expected to do in your own revision time. Add this to your revision schedule.

The Practice Book that accompanies this Revision Guide provides a fully developed Issue evaluation to help you become familiar with the revision process.

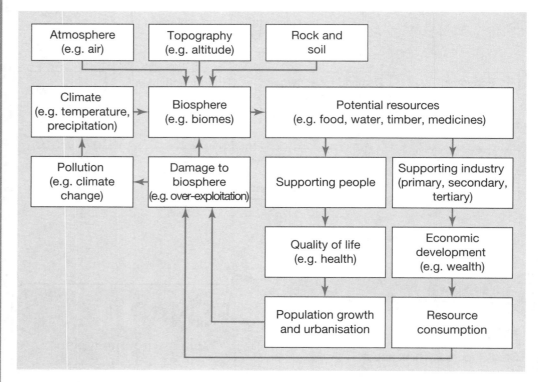

Figure 3 Living world and human world interactions

Possible issues

There is a range of possible geographical issues covered by the compulsory topics, and it would be a good idea to link these to real-world issues that happen in the time period between one and two years before the actual exam. Some possible issues include the following:

1. Managing issues linked to climate change such as: increased river flooding, more frequent droughts, stronger storms and sea-level rise.

2. Managing the reduction of carbon dioxide emissions such as: actions by individuals, families, organisations, businesses and governments.

3. Managing vulnerability to tectonic or weather hazards such as: adaptations, mitigation and resilience.

4. Managing changes to natural environments such as: controlling deforestation and protecting wilderness areas.

5. Managing urban infrastructure such as: transport, water, energy, sewerage systems and housing supply.

6. Managing urban regeneration such as: central business districts (CBDs), zones of discard and deindustrialisation areas.

7. Managing urban environments such as: air and noise pollution, traffic congestion, green spaces and use of brownfield or greenfield sites.

8. Managing the health and welfare of poor people in LICs such as: combatting disease, improving living conditions, reducing poverty and managing migration.

9. Managing the development gap such as: use of aid, foreign direct investment (FDI) and economic versus social development.

10. Managing environmental impacts of industrial activity such as: degradation of environment, derelict land and air and water pollution.

The resource booklet

It is important to know what the resource booklet will look like so that there are no surprises (also see the AQA GCSE Geography Exam Practice Book):

- There will be about six pages of figures, which will include information in a wide range of formats, including text, diagrams, maps, graphs, tables, photographs and so on.

- The information will be divided into several sections: (1) patterns and trends; (2) reasons for the patterns and trends; (3) challenges, issues, problems or pressures; and (4) possible solutions.

- Some figures may be included in the exam paper, ones where you have to complete a map or graph, for example, so you must consider and use the information from these in the exam as well as the ones given to you in the resource booklet.

- Make sure that you become familiar with all of the figures in the resource booklet before the exam, noting in particular where pieces of information support each other or contradict each other.

- Once you know the issue covered by the resource booklet you will know the theme of the exam; some additional research may be useful.

- You are not allowed to take any notes into the exam and you must have a clean copy of the resource booklet.

The Paper 3 Section A exam

It is important to know what the exam paper will look like so that you are well prepared (also see the AQA GCSE Geography Exam Practice Book):

- Issue evaluation appears as Section A and will have 37 marks.

- The time allowed for the whole exam (Sections A and B) is 75 minutes and there are 76 marks available (including six marks for SPaG), so you should work to a timescale of one minute for one mark. For example, a six-mark question should take you six minutes to complete. Leave at least nine minutes for the final question in Section A.

- There is no choice of questions; you must answer all of those set. It is best to do the questions in this exam in the order that they are set, as they are linked together and lead your thinking towards the final longer question.

- The types of question vary from one-mark skills questions, such as, *plot, complete* or *calculate* on figures provided in the exam paper; to multiple-choice (one mark), medium length answer questions that ask you to *assess, discuss* or *suggest* why (six marks); and one longer-answer *justify* question (nine marks), which also has an allocation of three SPaG marks.

- You should have a full range of stationery ready for this exam, including a calculator.

- Section B of the exam covers your fieldwork and will have 39 marks available.

The challenge of natural hazards

Draw a flow diagram, perhaps similar to Figures 1, 2 or 3, to show the issues associated with either tectonic or weather hazards.

Urban issues and challenges

Draw a flow diagram, perhaps similar to Figures 1, 2 or 3, to show the issues associated with either the urban world or urban change in the UK.

Justify

You should appreciate that there are several steps involved in a decision-making process. One of these is the ability to use geographical evidence and give geographical reasons for the decision or choice of option(s).

Academic skills required

- Use a wide range of geographical skills, such as:
    - cartographic (map interpretation)
    - graphical (graph interpretation and drawing)
    - numerical (e.g. using percentages)
    - statistical (e.g. calculating means)
    - qualitative data (using people's opinions)
    - quantitative data (using data facts).
- Use a well-developed academic writing style that shows an ability to use evidence, appraise advantages and disadvantages, evaluate, justify and provide conclusions.
- Recognise command words, such as **describe** or **compare**, and write answers that do what the command word says.
- Appreciate the opinions, attitudes and values of the people affected or involved with an issue (often called stakeholders).
- Identify and understand the links between the economic (money), social (people) and natural environments within geographical situations and issues.
- **Synoptic thinking:** make sure that you try to develop an ability to identify and understand the links between all of the different physical and human geography topics studied. Part of understanding an issue is to see how different geographical processes interlink to cause something to happen.
- **Decision making:** make sure that you are able to make geographical decisions. This means evaluating (looking for good and bad points) a range of possible options and choosing the best one based on the evidence available.

Raising the standard of your answers

- Follow the command word – answer the question in the way intended!
- Use evidence from the resource booklet to back up every point that you make. For example, evidence can be in the form of generalisations, quoting a short piece of text, quoting a fact or figure, getting information from a graph or map, or making reference to what a photo shows.
- In the longer six-mark questions, evidence should be taken from the figure stated in the question, but other resources may also be used. You may also need to show your own understanding of a topic.
- Use correct geographical terminology as much as possible.
- In the final nine-mark question, where you are asked to make a decision, you must remember that there is no correct answer. State your choice clearly in the space provided. Give the advantages of your choice with clear evidence quoted from the resources, but also recognise that your choice may still have some problems – again, make sure that you use evidence from the relevant resources. If you have time, you may wish to state the disadvantages of the other options, with the evidence from the resources. You *must* use evidence from the resource booklet and your own understanding throughout your answer.

NAILIT!

Evaluate

Issue evaluation means sorting through and understanding all of the information available on an issue and deciding what the positive (good) points are and what the negative (bad) points are. A higher step in this academic skill is to perhaps recognise that some points are more important than others, so there may be lots of small negatives, for example, but a couple of really important positives. Also, even the best option may have weaknesses, and the worst option strengths.

STRETCHIT!

Critical thinking

This is the process of thinking about a situation or issue and using facts to support an analysis of it, followed by an evaluation that leads to making a judgement about the situation or issue.

Example 6-mark question and answers:

Study Figure 4 below.

Suggest why people living in LICs/NEEs are more vulnerable to weather hazards than people living in HICs.

Use Figure 4 and your own understanding to support your answer.   [6 marks]

World Risk Index Report 2016
There is a very urgent need for action in Myanmar. This country, which is highly exposed to cyclones and floods, lacks a (stable) transport infrastructure. Freight transport can therefore easily collapse in the wake of an extreme natural event. The loadable electricity supply is poor on an international scale. In terms of 'logistics friendliness', the country is in the second-worst class. Japan is at the other end of the scale. This high-tech country is very highly exposed to natural hazards such as earthquakes and floods. However, thanks to its top values for all three indicators, it has very good prospects of mitigating a disaster resulting from such events.

Indicators for transport infrastructure, electricity supply and logistics friendliness

 Availability and quality of transport infrastructure: Extent of metalled roads, length of rail network, number of airports with metalled runways, container transshipment in ports measured in number of standard containers, per 100 000 inhabitants respectively; qualitative assessment of roads, rail infrastructure, ports and airports.

 Extent and quality of electricity supply: Percentage of population with access to electricity and quality of electricity supply.

 Logistic performance index: Logistic friendliness, measured in efficiency of customs and border clearance, the quality of the trade and transport infrastructure, the products of achieving competitive transport costs, the competence and quality of logistics service providers, the provision of shipment tracking facilities and the frequency of on-time deliveries.

Figure 4

NAILIT!

Sustainability

Make sure that you understand this term and can use the ideas in the Issue evaluation exam. Sustainability involves making sure that bad impacts are reduced or eliminated and good impacts are increased. This includes trying to make sure that, in the future, people and the natural environment do not experience a decline in the quality of life and, if possible, have a better quality of life.

NAILIT!

Empathy

Issue evaluation is very likely to involve the need to understand and appreciate the views and opinions of other individuals, groups or organisations. Make sure that you are able to put your own views and opinions aside, so that you can understand why these stakeholders see an issue in a certain way, and understand why conflicts arise.

Answer the six-mark question

Without looking at the two possible student answers shown, write your own six-mark answer to the exam question. In the exam you will have about 12 lines on which to write your response, so use this to judge the length of your answer. Try to write your answer in six minutes. After you have written your answer, compare it with the two provided and make a note of what you could have done to improve your answer.

Student answer A

Myanmar is highly exposed to cyclones and floods, but in terms of 'logistic friendliness' it is in the second worst class. Japan is a high-tech country, highly exposed to natural hazards such as floods; however, it has good prospects of mitigating a disaster. Myanmar was the second most affected country by extreme weather events between 1995 and 2014, such as Cyclone Giri, while Japan has a low climate risk index ranking. Only 40.6 per cent of Myanmar has access to electricity compared with Japan's 94.8 per cent, so there is no power after a weather disaster to help them cook and stay warm. The climate risk map shows that LICs/NEEs in tropical areas are most affected by extreme weather events, while HICs in cooler areas have lower risks.

Feedback

Student answer A shows a clear understanding of two countries, which are compared with some data from Figure 4. However, there is some close copying of text, which limits the student's expansion of application of knowledge and understanding. The answer does identify and explain some impacts and responses, but does not show understanding of the complex patterns.

Student answer B

The number of reported weather disasters is higher in the 21st century than in the 1980s and 1990s. For example, 325 storms and 550 floods in 2004–6, with damage costs highest in 2010–12 at over US$300 billion. The top ten countries most affected by extreme weather events during this time period are all LICs/NEEs and these have been unable to cope. For example, Myanmar has poor transport infrastructure, only about 41 per cent of people have electricity and there are poor trading links. These, along with the lack of other services such as health care, make the people vulnerable to extreme weather events, such as Cyclone Giri, and it takes them longer to recover from disease outbreaks. In contrast, Japan has a lot of weather hazards such as floods and typhoons, but a better infrastructure, external links and nearly 95 per cent have access to electricity. Being an HIC with access to technology and engineering makes people less vulnerable and allows protection measures and the ability to transport aid to those who need it. However, not all LICs/NEEs have a high climate risk ranking. Many African countries, for example, are in the least affected category, although global warming may change this in the future.

Feedback

In Student answer B, the student uses the full range of resources available, and includes the application of linked knowledge and understanding from Figure 4 and, in a couple of places, their own understanding beyond the weather hazard topic. There is an evaluation of the relative vulnerability of LICs/NEEs and HICs, and a comparison of two countries. This answer would gain higher marks in the exam.

CHECKIT!

The exam

1 In which exam paper and section is Issue evaluation?

2 a Approximately how many minutes per mark are available in this exam?

 b Are there skills questions in this exam?

 c Name the things that you need to remember about the final exam question in the Issue evaluation section of the exam.

The resource booklet

1 Describe the types of resource that could be present in the resource booklet.

2 Give the sections that the resources will be divided into.

3 Explain why you need to have an understanding of geographical ideas beyond the resource booklet.

Academic skills required

1 Give the method of calculating a mean.

2 What is an interquartile range?

3 Describe how you would complete a choropleth map.

Exam command words

1 Explain what is meant by the command word 'suggest'.

2 What does the command word 'plot' tell you to do?

3 Explain how you would justify a decision.

Fieldwork and geographical enquiries

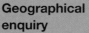

NAIL IT!

Geographical enquiry

This is the process of finding the answer to a geographical question or hypothesis.

Introduction

You will have carried out two geographical enquiries during your GCSE geography course that will have included the collection of primary data as part of a fieldwork exercise. The two enquiries will have been based on the physical and human geography elements that you have studied. They will have involved you investigating a geographical question that your teacher may have set you, for which you will have collected a range of data to help you answer the question.

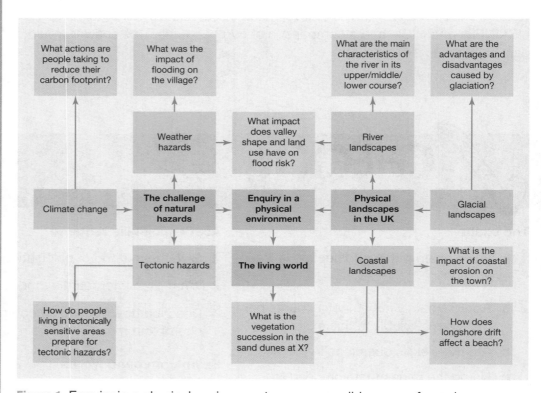

Figure 1 Enquiry in a physical environment – some possible areas of enquiry

DO IT!

Enquiry in a physical environment

Look at Figure 1. Use it to help you identify the fieldwork enquiry that you carried out in a physical environment, and identify which topic it relates to. Make sure that you know the title of the fieldwork enquiry, as you will need to recall this in the exam.

Fieldwork in a physical environment

Complete a mind map showing all the ways in which you collected fieldwork data in a physical environment to help you answer your enquiry question. Try to recall the process of data collection that you went through and record how you did it.

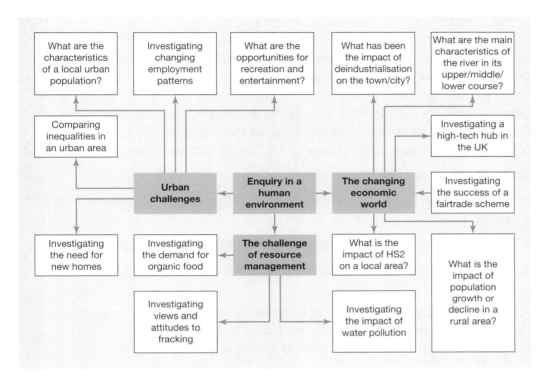

Figure 2 Enquiry in a human environment – some possible areas of enquiry

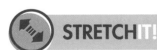

Fieldwork in a human environment

Complete a mind map showing all the ways in which you collected fieldwork data in a human environment to help you answer your enquiry question. Try to recall the process of data collection that you went through and record how you did it.

SNAP**IT!** Fieldwork enquiries

Snap an image of each of your mind maps. Discuss these with a friend to describe the ways in which you collected your fieldwork data in both a physical environment and a human environment.

DOIT!

Enquiry in a human environment

Look at Figure 2. Use it to help you identify the fieldwork enquiry that you carried out in a human environment, and identify which topic it relates to. Make sure that you know the title of the fieldwork enquiry, as you will need to recall this in the exam.

How to revise for Paper 3 Section B

Revision and preparation for the fieldwork questions will be different from Units 3.1 and 3.2. Papers 1 and 2 are based on the physical and human geography topics, but Section B of Paper 3 (Fieldwork) is broken down into two sections:

- questions based on the use of fieldwork materials from an unfamiliar context
- questions based on the two pieces of individual enquiry work that you carried out.

In the fieldwork section of the exam you will, therefore, need to make sure that you know the *techniques* that can be used to collect geographical data through fieldwork and the *methods* that can be used to present geographical data. You will also be expected to interpret and explain maps, diagrams and geographical data. Your main knowledge and understanding revision for Paper 3 will be completed when you revise the topics within 'Living with the physical environment' and 'Challenges in the human environment'.

What you need to concentrate on is:

- your ability to answer the style of question that appears in the fieldwork section of Paper 3
- developing familiarity with the process of geographical enquiry
- making sure that you can explain and interpret a range of geographical data to reach conclusions.

The AQA GCSE Geography Exam Practice Book that accompanies this Revision Guide provides a set of questions based on the fieldwork exam questions to help you become familiar with the revision process.

The process of enquiry

There are six strands of fieldwork enquiry that you can be asked about in the exam. The questions could relate to fieldwork enquiries that you carried out, or to data relating from unfamiliar enquiries that the examiner has included in the exam paper. It is important that you understand each of the six strands clearly so that you know which aspect of the fieldwork you are being tested on. The six strands are as follows:

1 Geographical enquiry

This is the selection of a suitable geographical question that you can set out to answer. You will need to know the titles of the two geographical enquiries that you carried out in both a physical environment and a human environment, as you will be expected to write these out as part of the exam. You also need to be able to justify why it is a suitable enquiry and how each question relates to an area of geography that you have studied as part of the GCSE course. Finally, it is important that you are aware of the potential **risks** of carrying out fieldwork activities and how these risks might be reduced – this could be in the form of a simple risk assessment.

DO IT!

Paper 3 revision schedule

Look back at the two pieces of fieldwork that you completed during your GCSE geography course. Make sure that you include time in your revision schedule to focus on these and practise your geographical skills.

2 Measuring and recording data

There are two types of data that can be used in your geographical enquiry: primary data and secondary data.

- Primary data collection is where you physically collect the data yourself – this could be measuring the width of a river, or counting the number of pedestrians on a street, for example.

- Secondary data is where somebody else has collected the data – this could be using data from a website or reading an article in a newspaper.

You need to be able to show that you understand how you could collect appropriate physical and human data, as well as being able to describe the actual process of how the data is collected and the sampling methods that you may have used.

3 Processing and presenting fieldwork data

For this part of the geographical enquiry, you need to be aware that there is a wide range of cartographical and graphical methods that can be used to present your data, including maps, line charts, bar charts, pie charts, pictograms, histograms with equal class intervals, divided bar charts, scattergraphs, and so on. You will need to be able to draw and complete maps and graphs from data that is provided in the exam and be able to write about selecting appropriate data presentation methods.

4 Describing, analysing and explaining fieldwork data

A key part of your geographical enquiries will be making sense of the data that you collected. This is the process of summarising the data that you have collected, then carrying out an analysis of the data followed by some explanations of the results of that fieldwork data. You may have also carried out some statistical analysis of the data that you collected, perhaps using the median or mean, or by calculating quartiles and the interquartile range, for example.

5 Reaching conclusions

This is the part of the geographical enquiry that is often most overlooked. It involves referring back to the original enquiry question that was set and producing a brief summary of the results of your data analysis in order to answer the question set. This can often be in the form of a short paragraph that gives a concisely written response to the enquiry question.

NAILIT!

Data presentation

Look back through the two fieldwork enquiries you carried out. Ensure you understand the data presentation methods you used in both. Make sure that you know why you used those methods of data presentation so that you could explain this in the exam.

NAILIT!

Results and conclusions

For both of your fieldwork enquiries you have worked on, make sure that you have an analysis of the results of the data that you collected and that you know what it means. It really helps to have produced a range of maps and graphs that you can refer back to in the exam. Next, draft a written conclusion to both of your enquiries, so that you know the answers to the questions that are being investigated.

6 Evaluation of geographical enquiry

Finally, you will need to be aware of the strengths and weaknesses of your enquiries and how they could be improved. There are four aspects of this section that you need to be able to evaluate:

- Any problems that you may have had collecting your data.
- The limitations of the data that you collected.
- Suggestions of any other data that you could have collected.
- How reliable your conclusions were.

DO IT!

Your fieldwork

Work through both of the fieldwork enquiries that you carried out and create a table to record the work that you did for each of the six strands of fieldwork enquiry. Remember that you will need to do this for both of your fieldwork enquiries – one in a physical environment and one in a human environment.

The Paper 3 Section B exam

It is important to know what the exam paper will look like so that you are well prepared (also see the AQA GCSE Geography Exam Practice Book):

- Fieldwork appears as Section B in Paper 3 and has 39 marks available including SPaG.

- The time allowed for the whole exam is 75 minutes and there are 76 marks available, so you should work to a timescale of one minute for one mark. For example, a six-mark question should take you six minutes to complete. Leave at least nine minutes for the final question in Section B.

- There is no choice of questions; you must answer all of those set. The first questions in this section will relate to fieldwork in general, while the questions at the end will focus on the two fieldwork enquiries that you carried out. It does not matter which order you do them in, although working through the questions on your enquiries in order will help to develop your thinking towards the final longer question.

- The types of question vary from one to two mark skills questions, such as *plot*, *complete* or *calculate* on figures provided in the exam paper, to short-answer questions that may ask you to *describe* or *justify* (three to four marks); medium length answer questions that ask you to *assess*, *discuss* or *suggest why* (six marks); and one longer answer *explain* question (nine marks), which also has an allocation of three SPaG marks.

- You should have a full range of stationery ready for this exam, including a calculator.

NAIL IT!

Data response questions

You should be aware that you will be expected to respond to fieldwork data provided in the exam paper. Therefore, you need to be familiar with a range of data presentation techniques and be able to both describe and explain what the data shows.

Academic skills required

The things that you need to be able to do are:

- Use a range of geographical fieldwork techniques to investigate questions and issues in relation to geographical enquiry.

- Use a wide range of geographical skills, such as:

  - cartographic (map interpretation)

  - graphical (graph interpretation and drawing), numerical (e.g. using percentages)

  - statistical (e.g. calculating means)

  - qualitative data (using people's opinions)

  - quantitative data (using numerical data as facts).

- Use a well-developed academic writing style that shows an ability to interpret, analyse and evaluate information, communicate findings, justify and provide conclusions in relation to geographical enquiry.

- Recognise command words and write answers that do what the command word says.

STRETCHIT!

Drawing conclusions from fieldwork data

This is the process of interpreting the data that you collected for your two fieldwork enquiries. You need to be able to relate it to the question you originally set and produce a written statement that summarises your findings.

Raising the standard of your answers

- Follow the command word—answer the question in the way that is intended!

- Complete the fieldwork data questions as accurately as possible – make sure that you have the right equipment to be able to do this in the exam. You may be required to do calculations as well.

- Respond to the data provided when asked. This might mean that you need to describe any patterns in the data, explain what they mean, or even suggest alternative ways of presenting the data.

- In the longer six-mark and nine-mark questions, you will need to answer based on the two fieldwork enquiries that you carried out. You could be asked questions about any of the six strands of fieldwork enquiry. For these questions, you will always be asked to state the title of your fieldwork enquiry. You *must* use evidence from your fieldwork enquiries and your own understanding throughout your answers.

Example nine-mark question and answers:

For one of your geographical enquiries, evaluate the success and reliability of your data collection methods.

Give the title of your fieldwork enquiry. [9 marks]

Student answer A

Title of fieldwork enquiry: How does a beach change?

Throughout my fieldwork, I got some very accurate results and there was a lot of evidence to explain my results in my conclusion. The problem I had getting the results were that they were all similar to each other. This made it hard to get an average overall. If I repeated the work, I would manage my time better and not rush collecting my data on the beach. I would go back again for another day. The strength is that I collected the data accurately using the ranging poles. I could ensure that the data is more reliable if I repeated my tests more than once to make sure that they were accurate.

Feedback

Student answer A above shows a vague recollection of collecting data on a beach and some understanding of the fact that the data may have been reliable. Unfortunately, there is very little reference back to the question and little explanation of the success of the enquiry. It highlights the fact that you need to know all the six strands of your fieldwork enquiry; if you don't, it is very difficult to answer these questions well.

Student answer B

Title of fieldwork enquiry: To what extent does the beach profile at Newhaven change from west to east?

The beach profile was measured from shore to sea at 10 locations 100 metres apart along Newhaven beach. This produced a large amount of data for 1 km along the beach, which was enough distance to see if there was any change. The sizes of beach material were also measured at each profile, which helped to make links between the angles and pebble sizes which helped to explain the changes. A low tide time was chosen so that the lower beach could be measured as well as the upper beach, this gave a complete profile. There were some weaknesses, for example, the whole length of the beach could have been measured to provide even more data to increase the reliability of the results. Each of the 10 profiles was measured by different people and this may have increased human error, it would have been better for the same group to take all of the measurements. Some groups used a 'gun clinometer' which is not as accurate as the rectangular clinometer which could be placed directly on the beach. Reliability could also have been increased by measuring profiles before and after a storm to see the effect of different types of wave on the beach profile.

Feedback

Student answer B focuses on the two key elements of the question; that is, the *success* and *reliability* of the data collection methods. Firstly, it is evident that this student knows the title of the enquiry question that they set out to answer. There is then a clear reference to the fact that the data collection was successful because it was possible to answer the enquiry question. This is followed by a comment on the reliability of the data and how the data collection could have been improved. This answer would gain the better mark in the exam.

 # CHECKIT!

The exam

1 Which exam paper and section is the fieldwork enquiry found in?

2 a How many minutes per mark are available in this exam?

  b Are there skills questions in this exam?

  c Name the things that you need to remember about the final exam based on your geographical enquiries.

The enquiry process

1 Describe the locations of both of your fieldwork enquiries.

2 State the six strands of enquiry that you could be examined on.

3 Explain how your geographical enquiries relate to the topics you have been taught.

4 What risk assessment did you carry out for your fieldwork enquiry?

Academic skills required

1 Explain the method of calculating the interquartile range.

2 Describe how you would complete an isoline map.

3 Explain how you would justify data presentation methods.

Exam command words

1 Explain what is meant by the command word 'suggest'.

2 What does the command word 'calculate' tell you to do?

3 What does the command word 'explain' tell you to do?

4 What does the command word 'assess' tell you to do?

Glossary

abiotic   the non-living part of an ecosystem, such as soil

abrasion   the wearing away of cliffs by sediment thrown by breaking waves

abstraction   the removal of water from the ground, rivers or lakes so that it can be used by people

accessibility   how easy it is to get to and from a place

active layer   the seasonally thawed surface layer above permafrost

adult literacy rate   the number of adults in a country who can read and write (usually expressed as a percentage)

afforestation   planting of trees on a large scale

agribusiness   a large farm or group of farms organised and managed efficiently to make as much profit as possible

aid   the giving of money, expertise or technology by one country to another to help development

appropriate technology   a level of equipment and machinery that people can understand easily and does not cost too much (see *intermediate technology*)

aquifer   a layer of rock that contains groundwater

arête   sharp razor like ridge formed between two corries

attrition   erosion caused by rocks transported by waves that bump into each other and break into smaller pieces

autotroph   a plant that uses sunlight, nutrients and water to grow (sometimes called producers)

basal slip   a glacier sliding over the bed due to meltwater under the ice acting as a lubricant

biodiversity   the number and variety of species found within an ecosystem

biomass   the weight of living matter (all the plants and animals) in a given area; organic matter used as fuel in a power station

biome   a large-scale global ecosystem, such as a tropical rainforest (see large-scale global ecosystem)

biotic   the living part of an ecosystem, such as plants

birth rate   the number of live births in a year within a population of an area (usually expressed out of one thousand people)

boulder clay   clay containing many large stones and boulders, formed by deposition from melting glaciers and ice sheets.

bulldozing   when ice in a glacier pushes loose material in front of it

canopy   a layer of trees within a forest ecosystem

carbon footprint   the amount of carbon dioxide produced by a person, household or business

climate   the long-term patterns of average precipitation and temperature, including hot and cold seasons or wet and dry seasons, for an area

climate change   a significant change in the expected long-term patterns of average precipitation and temperature for an area; this may be a natural change or due to human activities

colonial expansion   the historical takeover of overseas territories by a powerful country, usually for economic benefit

commercial farming   the growing of crops or rearing of livestock to make money

constructive wave   waves that build up beaches by pushing sand and pebbles further up the beach

consumer   a person, or group of people, who buys or uses things; or in an ecosystem, a creature that eats plants or other creatures

container port   a place where ships designed to carry large metal containers can dock to load or offload cargo. Containers are a standard size so that they can be transferred from ships to rail or lorries

corrasion   acids contained in sea water will dissolve some types of rock such as chalk or limestone

corrie   hollow formed in a mountainside by glacial erosion, rotational slip and freeze-thaw weathering. It is where a glacier begins. A small circular lake called a tarn is left when the ice melts

cycle of poverty   a situation where poor people become stuck in poverty with no way of changing their conditions

cyclone   tropical storms that occur in South-East Asia

death rate   the number of people who die in a year within a population of an area (usually expressed out of one thousand people)

debt reduction   also known as 'debt-for-nature swap', occurs when there is an agreement between a country that owes a huge amount of money internationally and a lender such as the World Bank, where some of the debt is wiped away in return for legal protection of part of an ecosystem

decomposer   a living organism, such as fungi or beetles, that helps the breakdown of dead living matter

deforestation   the cutting down of trees on a large scale

de-industrialisation   a stage in economic development where manufacturing industries decline and close down, while service industries grow

democracy   a political system where people have a free vote in elections and the right to free speech

Demographic Transition Model (DTM)   a graph representing changes in the population of a country or region, by tracking birth and death rates and population size over a long period of time

depopulation   a decline in the number of people in an area, especially due to movements away (emigration)

desalination   a process of creating fresh water from salt water

desertification   the change in a semi-arid area to a desert because of a change in natural processes or damaging human activities

destructive wave   waves that erode beaches with a strong backwash that removes beach material

diplomatic link   a way of countries communicating with each other to ensure understanding

discharge   the volume of water flowing in a river

diurnal temperature range   the daily change in temperature between day and night

drainage basin   an area of land drained by a river and its tributaries

drought   a situation where the fresh water supplies are well below the amount needed to support the population of an area

drumlin   a hill made of glacial till deposited by a moving glacier, usually elongated or oval in shape

ecological footprint   the amount of resources used from the Earth to support a population

economic development   the creation of industries and businesses in a country or region in order to make money

ecotourist   a person who goes on holidays to appreciate and look after nature; a type of holiday that causes little or no damage to the natural environment

emergent tree   a very tall tree that is above the canopy layer

energy gap   the difference between the amount of energy available for a country and the amount of energy needed

energy mix   the variety of sources of energy production of a country

enhanced greenhouse effect   the warming of the atmosphere caused by human activities, extra to the warming caused by natural processes

environmental sustainability   the long-term balance of natural systems

erosion   wearing away or removal of material by a moving force such as water

erratic   rocks transported and dumped by glacial ice to a different location

evapotranspiration   the process where plants give off water through their leaves and then this water is evaporated from the leaf surface into the atmosphere

famine   a situation where there is not enough food for the population of a place

favela   a squatter settlement in a city of Brazil

fetch   the distance over which a wave can travel uninterrupted

floodplain   flat area of land forming the valley floor either side of the river channel

food chain   a simple sequence of links between living things within an ecosystem, where one living thing eats another

food insecurity   a situation where the food supplies for an area are decreasing

food miles   the distance over which food is transported from where it was farmed to where it is eaten

food security   a situation where the food supplies are enough for the population of an area

food web   a complex series of links between living things within an ecosystem where energy is transferred through food

Foreign direct investment (FDI)   when businesses from one country invest money in the businesses and industry of another country

formal economy   businesses and industries that offer regular wages and contracts, with a set place of work or work routines, which follow laws and pay taxes

fossil fuel   an energy source that was created by geological processes, such as oil

fragile environment   a natural area and its systems that can be easily damaged by even slight changes

freeze-thaw weathering   when water seeps into cracks in rocks, freezes and expands, eventually breaking the rock apart

geological process   a process linked to the formation of rocks, either sedimentary, igneous or metamorphic

geothermal energy   a source of heat from very hot magma underground that can be used to turn water into steam and create cheap electricity or provide heating to homes and settlements

glacial period   an interval of time (thousands of years) within an ice age that is marked by colder temperatures and glacier advances

glacier   a slowly moving mass or river of ice formed by the accumulation and compaction of snow on mountains or near the poles

global atmospheric circulation   the large-scale pattern of movement of air in the atmosphere, such as convection cells and trade winds

global ecosystem   a large-scale community of plants and animals adapted to the physical conditions of the region, found in several regions of the Earth

globalisation   the linking of people and countries all over the world by various processes such as communications, trade, migration, money and culture

government policy   the approach of a country towards situations and processes

green technology   machinery and other technologies that use natural systems as a resource without damaging them

greenhouse effect   the trapping of heat within the atmosphere by denser gases, such as carbon dioxide

greenhouse gas   one of the denser gasses responsible for trapping heat in the atmosphere, such as methane or carbon dioxide

growing season   the number of months in a year when the average temperature is high enough for plants to grow

Hadley convection cell   one of the main large-scale movements of air in the atmosphere, found either side of the equator all around the planet

hard engineering   the building of entirely artificial structures using various materials such as rock, concrete and steel to reduce, disrupt or stop the impact of river or coastal erosion

high income country (HIC)   a country that is very wealthy compared with other countries based on World Bank income classification, sometimes called 'developed'

Human Development Index (HDI)   a measure of how developed a country is using social and economic indicators

human rights   people have the right to experience life without persecution from others, this is supported through the United Nations

humanitarian crisis   a situation where the lives of many people are threatened by a natural or man-made hazard

hurricane   a tropical storm located in the Atlantic (in the Pacific they are known as typhoons)

hydraulic action   the force of the river against the banks causing air to be trapped in cracks and crevices. The pressure weakens the banks and gradually wears it away

hydro-electric power (HEP)   a source of electricity created by water flowing through turbines

hygiene   the absence of germs through cleanliness and understanding of how to reduce illness

indigenous   native to an area

infant mortality   the number of deaths of children under the age of one year, usually expressed out of one thousand live births

informal sector the part of the economy that is not official, often involving no fixed place of work or regular income with no taxes paid and no welfare plan

infrastructure the structures and things that are needed to support businesses, industries or people, such as transport networks, power and water supplies

integrated transport network the linking together of different methods of transport, perhaps with stations close to each other

inter-glacial period a longer period of warmer conditions between glacial periods

intermediate technology a level of machinery and equipment that is above a simple level (hand tools) but below the complicated level (high-tech), usually more affordable and easier to understand (see *appropriate technology*)

international aid help given by a country or international organisation to a poorer country or one in need of assistance

intertropical convergence zone (ITCZ) the linear area where the two Hadley cells meet near the equator; it moves according to the season, for example, in the northern hemisphere summer it is north of the equator but in winter it is south of the equator

large-scale global ecosystem an ecosystem that is found in several continents and countries around the world, usually at a similar latitude (see *biome*)

latitude the distance, north or south, from the equator (based on an angle)

leaching the loss of water-soluble plant nutrients from the soil

levée raised bank found on either side of a river, formed naturally by regular flooding, or man-made as a flood defence

life expectancy the average number of years from birth that a person living in a place can be expected to live

location place where something is found in relation to other important places or features

longitude the distance east or west from the Greenwich meridian line (based on an angle)

longshore drift transport of sediment along a stretch of coastline caused by waves approaching the beach at an angle

low income country (LIC) a country that is poor compared to other countries, sometimes called less developed; based on World Bank income classification

magma molten rock below the Earth's surface

malnourished when people do not have a balanced diet

managed retreat controlled retreat of the coastline, allowing flooding to occur over low-lying land

microfinance loan the lending of a small amount of money to poor people so that they can make significant improvements to their lives and work in times of hardship

migration the permanent movement of people from one place to another for at least one year

moraine frost shattered rock debris and material eroded from the valley floor and sides, transported and deposited by glaciers

nationalised when a government takes over a business or industry

natural hazard a natural event (for example, an earthquake, volcanic eruption, tropical storm, flood) that has the potential to cause damage, destruction and death

natural increase the increase in the population size of an area due to the birth rate being higher than the death rate

natural resource a product from natural systems, such as ecosystems or geological processes, that is of use to people

newly emerging economy (NEE) a country that has recently started to create wealth through developing businesses, industries and trade; sometimes described as developing or an NIC

newly industrialised country (NIC) a country that has concentrated investment on secondary industries in order to make money through exporting to other countries; sometimes called 'developing' or NEE

non-government organisation (NGO) an organisation that is not run by a government, but independent such as a charity

nutrient a mineral that can be used by a living organism to help it grow and develop

nutrient cycle the way in which minerals are stored and moved around within an ecosystem

organic derived from living organisms

outwash material carried away from a glacier by meltwater and deposited beyond the moraine

over-cultivation growing too many crops in an area so that the soils are seriously damaged and lose their nutrients

overgrazing rearing too many animals in an area so that the vegetation and soils are seriously damaged

people per doctor a way of understanding the level of health care in a place, the fewer people per doctor, the better the health care should be

permafrost the ground that is frozen all year except for a thin upper layer in summer

photosynthesis the process plants use to grow by using sunlight, carbon dioxide and water

polar region an area of the Earth near either the North or the South Pole

post-industrial a period of time when the number of secondary industries has declined to be replaced by tertiary businesses

precipitation all types of water falling from the air, such as rain, snow, sleet, or hail

preferential aid help that is specially given from one country to another, maybe to strengthen political support

primary producer a plant that is at the base of a food web or trophic levels, which is able to use sunlight to grow (using the process of photosynthesis)

privatisation where nationalised businesses and industries are 'sold off' by a government to be run separately without government interference

pyramidal peak where several corries cut back to meet at a central point, the mountain takes the form of a steep pyramid

quality of life a way of considering the well-being of people by looking at a wide range of factors linked to health, housing, employment, clean water, food supply, education and political freedom, for example

Quaternary period the last 2.6 million years during which there have been many glacial periods

quaternary sector industries providing information services, such as computing, ICT, consultancy (business advice) and research and development

recession an economic situation where businesses and industries find it difficult to sell things and so everyone is poorer; on a world scale this has happened about every 50 years

refugee a person fleeing from one country to another to escape the threat of death

regeneration investment of money, infrastructure or business in a place to help it recover from a decline

relief the height of land

remittance an amount of money sent home to families by migrants working in another country

renewable energy a source of power that is not using fossil fuels but instead uses something that will not run out, such as wind or sunlight

research and development the way in which businesses and industries stay ahead of their rivals by investigating new ideas and products

resource (see *natural resource*)

rotational slip occurs when the ice moves in a circular motion. This process can help to erode hollows in the landscape and deepen hollows into bowl shapes

rural–urban fringe the zone around a large settlement which has a mixture of urban and rural land uses and a mixture or rural and urban processes

rural–urban migration the movement of people from rural areas into towns and cities, with a permanent change of residence

safe water clean fresh water that is free of contamination or harmful bacteria

semi-arid dry areas of the world, with a higher total rainfall than deserts but not enough to support a biodiverse ecosystem or lots of people

service sector also known as the tertiary sector, is the collection of businesses based on providing a service for people or other businesses, such as banking and insurance

shrub layer a low level of vegetation in an ecosystem, below the canopy layer but above the ground layer

small-scale local ecosystem a balanced community of plants and animals living within a small area

snout 'nose' or end of the glacier

soft engineering involves the use of the natural environment surrounding a river, using schemes that work with the sea or river's natural processes.

solar energy ways of collecting and using the power of the Sun, for example, through solar panels to create electricity

solifluction   the gradual movement of wet soil or other material down a slope

spending on education and health   the quality of life of a population often depends on their health and being able to read and write, so governments invest money in these services

starvation   when people do not have enough food to eat

striation   long, straight, parallel lines or grooves in a bedrock surface, formed by boulders, gravel and pebbles embedded in a glacier that has passed over the surface

subsistence farming   growing of food and rearing of animals just to feed a family. Sometimes surpluses may be sold

sustainable   ensuring that a situation or process can continue into the foreseeable future without damaging the natural environment

thalweg   the line of fastest flow along the course of a river

till   sediment deposited by melting glaciers

trading   the selling or exchange of goods and services, usually between countries

transnational corporation (TNC)   a large international business that has factories or offices in several countries around the world

transpiration   the process of plants giving off water through their leaves

trophic level   the arrangement of 'feeding groups' within any ecosystem, from plants at the base to top carnivores at the peak

tundra   a cold semi-arid environment with very cold temperatures for most of the year and only a sparse vegetation of lichens, grasses and small shrubs

typhoon   the name given to a tropical storm in the Pacific (in the Atlantic they are known as hurricanes)

undernutrition   where people do not have enough variety of minerals and vitamins in their diet

urban sprawl   the spread of urban land uses and influences into a rural area surrounding a city or large town

urbanisation   the rise in the percentage of people living in an urban area compared to rural areas

water cycle   the ways in which water is stored and moved on the planet

water deficit   a situation where the amount of fresh water available is below the amount that is needed

water stress   a situation where the abstraction of fresh water is greater than the rate of replacement, perhaps because of overuse or a drier climate

water surplus   a situation where the amount of fresh water available is above the amount that is needed

'Western' culture   a way of living; applied to countries whose population is historically largely based on European immigration, such as the USA, Australia or Europe itself. The model is that of freedom, wealth and the ability of any person to be successful

wilderness   an area of the Earth that has not been altered by human activity, or has only experienced minimum human interference

wind energy   ways of collecting and using the movement of the air, for example, by windmills or modern turbines in wind farms to create energy

Answers

For answers to all Check it! questions, visit: www.scholastic.co.uk/gcse

Natural hazards: Review It! (p. 30)

1 A natural hazard is a naturally occurring physical phenomena, which poses a potential risk to human life and/or damage to property

2 The effects of a tropical storm can be reduced by: monitoring the weather through satellites to provide early warnings and protection through reinforcing windows or building houses on stilts to protect against flooding from storm surges.

3 Earthquakes and volcanoes mainly occur along plate margins.

4 The benefits of living near a volcano are fertile soil for farming, jobs in the mining industry and getting energy through geothermal power.

5 Tropical storms are formed through the following processes; firstly the air is heated above the warm tropical oceans, causing air to rise rapidly. This upward movement of the air draws up water vapour from the ocean's surface. The evaporated air cools as it rises, which causes it to condense to form large thunderstorm clouds. As the air condenses it releases heat, which powers the storm and causes more and more water to be drawn up from the ocean. Several thunderstorms can join together to form a giant spinning storm. It will officially be classified as a storm when winds reach 63km/h. As the storm moves across the ocean it develops in strength and then when the storm hits land it loses its momentum as friction with the land causes it to slow down and weaken.

6 Following a natural disaster there will be immediate, short- and long-term responses. Long-term responses may be seen to be more significant because they will then reduce the impacts of a natural disaster in the future. For example, following the Nepal earthquake 2015, the buildings were built to stricter codes so they would be less likely to be collapse in a future earthquake. Countries that are more prepared are often quick to respond to a disaster and they will have planned long term in case of a natural hazard. For example, after the L'Aquila earthquake in 2009, the DEC (Disasters Emergency Committee) did not need to provide any aid because Italy is a developed country. In a developing country, immediate responses are more significant than long-term responses, because the country may not be able to afford long-term planning, such as, hazard-resistant buildings. For example, after the Nepal earthquake in 2015, the DEC raised US$126 million to help support the country's redevelopment.

7 When two plates meet, the denser oceanic plate is subducted beneath the less dense continental plate. As the oceanic plate moves downwards it melts and this creates magma, which is less fluid than at a constructive margin. The magma can break through to the surface to form a steep-sided composite volcano. Eruptions are often very violent and explosive.

8 There are cyclical changes in solar energy outputs linked to sunspots. A sunspot is a dark patch on the surface of the Sun. The number of sunspots increases from a minimum to a maximum over a period of 11 years. The more sunspots there are, the more heat that is given off from the Sun, which can lead to higher temperatures on Earth.

9 Carbon dioxide accounts of approximately 60% of enhanced greenhouse emissions and concentrations have increased by 30% since 1840. Concentrations have increased due to the burning of fossil fuels in industry and power stations. Higher concentrations of carbon dioxide are also caused by transport, such as car exhausts, and due to deforestation.

10 There were a range of primary and secondary effects of the 2015 earthquakes in Nepal. A primary effect was the amount of people that were killed and injured in the earthquakes. In total 9000 people died and 22,000 people were injured. These levels were high as the earthquake happened in a developing country, which meant that buildings were not constructed to strict codes and so a large number of buildings were destroyed. Secondary effects included avalanches caused by the earthquake, including one on Mount Everest where 21 people were killed and one in the Langtang region which resulted in 250 people missing. Communication links were disrupted with landslides blocking roads. Overall, the earthquake had an extremely big impact on Nepal and cost US$5 billion in damage.

11 The 2014 Somerset floods in South West England caused a range of effects. Social damage to buildings, including 600 homes flooded and 16 farms evacuated, resulting in a large amount of people requiring temporary shelter. Economic: it caused £10 million worth of damage

and approximately 14,000 ha of farmland were flooded, which affected farmers' livelihoods. Environmental: sewage contaminating the flood water and stagnant floodwater had to be re-oxygenated before being pumped back into the river.

12 There is evidence to support or to reject this statement. Comparisons and conclusions are difficult, due to the different magnitudes, timings and depths of different earthquakes. An HIC will have a more developed infrastructure than an LIC or NEE, which will be expensive to replace. The L'Aquila earthquake in Italy in 2009, for example, cost US$16 billion but the earthquakes in Nepal in 2015 cost US$10 billion, although the earthquakes in Nepal were of a higher magnitude. More developed countries may have made adaptions to buildings, or they may be built to better building laws, resulting in fewer deaths and injuries. In the Nepal earthquakes, 9000 people died and 20 000 people were injured, compared to the 6.3 magnitude earthquake in L'Aquila in Italy, where 308 people died and 1500 people were injured. However, the number of deaths may also be related to the level of development of a country, as HICs usually receive aid very quickly after an earthquake but it may take days for remote areas in LICs to receive aid due to damage to the communications infrastructure. The economic effects in a developed country will be higher but the social effects are generally more devastating in a developing country.

The living world: Review it! (p. 59/60)

Ecosystems

1 a One of: rock type, soil characteristics, or amount of water.

   b One of: climate (rainfall, temperature, seasons), linked to distance from equator, oceans or mountain ranges.

2 Answers will vary according to case study. Here the example is Epping Forest, Essex.

   a Two, such as: English lowland wood; pollarded trees with many thick branches; deciduous trees such as oak, beech, birch; 2500 ha in area; rare and vulnerable fungi; consumers such as grey squirrels and muntjac deer.

   b One of: enclosing the forest prevents the deer from roaming and causing damage to the forest; The Epping Forest Act stopped pollarding but then the trees blocked out light from the ground level vegetation; busy roads around the forest kill deer; human recreation such as walking, horse riding and mountain biking disturb and damage wildlife; air pollution has affected the older trees.

   c One of: powerful storms, such as the Great Storm of 1987; occasional droughts, such as 1976–77; a large deer population may lead to over grazing and damage to the forest structure.

Tropical rainforests

1 a The structure has three main layers: emergent (50 m tall), main canopy (35 m) and sub-canopy. The shrub and ground layers are not significant.

   b The canopy layer, which consists of evergreen trees, blocks sunlight from reaching the ground layer. Without sunlight the vegetation cannot photosynthesise and grow.

2 a Most creatures live in the canopy layer.

   b This is where most of the food is (fruits, nuts, nectar). The ground level is dark and plants do not grow well so there is little food at this level.

3 There is a wide range of adaptations as this is a very old ecosystem, relatively unaffected by world climate cycles, and so creatures have had a long time to evolve. Ideal growing conditions (hot, wet, sunlight) have enabled plants and animals to adapt to a balance with the elements and also with each other.

4 Any one of: extinction of species; disruption of food chains and webs; loss of genetic material which could have been useful to people; the ecosystem becomes more fragile and less resistant to change (e.g. in climate).

5 a The soil is exposed to erosion, possible local flooding, natural vegetation unable to regrow and loss of local biodiversity.

   b The absorption of CO_2 is greatly reduced, so it is not stored and more remains in the atmosphere, increasing the 'greenhouse effect'. Loss of global diversity. Loss of potential health products/cures.

6 Indigenous populations have a smaller impact than those with strong economic motives. Tribes make their homes from timber and clear small patches of land for crops, using only what they need to survive and affecting only a small area. Those with economic motives have access to machinery which can destroy large areas quickly, especially with open cast mining, which strips away large areas of forest, or reservoirs that flood large areas. The World Bank and large companies invest in developing countries to build dams for the generation of electricity.

7 Answers will vary according to case study. Here the example is Malaysia in South-East Asia.

  a Four causes of deforestation: logging of tropical hardwoods to export, bringing money into the country; cheap electricity needed for businesses and industries so HEP stations with dams have been constructed; mining of tin and extraction of oil and gas to provide resources for the country and export income; palm oil plantations to create an export product and income; population growth means that living space is needed for poorer people, who can use the area.

  b Impacts: reduction of biodiversity; wildlife endangered (e.g. orang-utans) by loss of habitats; soils exposed to erosion; transpiration reduced; absorption of carbon dioxide reduced.

  c Impacts on people: jobs created in mining, farming, energy, logging; export earnings from selling rainforest related products abroad; water and air pollution increased by human activities in rainforest; soils degraded to a point where they will no longer produce crops; climate change caused by upsetting the carbon dioxide and oxygen balance in atmosphere; more extreme climate events – floods, droughts.

8 a The rainforests can provide medicines to help improve health, they provide important food and timber resources which could be managed sustainably. The loss of biodiversity could disturb natural balances and cause losses of invaluable natural products and extinction of species. The rainforest is believed to help balance the Earth's atmosphere, and to help balance oxygen and carbon dioxide within the atmosphere, which may be important to reducing climate change.

  b Any sustainable strategy, such as international agreements: most countries are linked through globalisation processes, and communications such as the internet can help educate and persuade all of those in LICs/NEEs and HICs of the need to aim for sustainability. HICs can provide monitoring of rainforest areas (e.g. satellite technology) so that it is known how much is being cut down and whether protected areas are being damaged, and provide funding through schemes such as REDD, which emphasises management steps, such as stopping deforestation, using forests sustainably, recovering degraded land and expanding forest cover.

Hot deserts

1 a It is the difference between the maximum daytime temperature and the minimum night-time temperature in a 24 hour period.

  b The diurnal temperature range is large, with, for example, a high of 38°C and a low of 5°C, giving a diurnal range of 33°C.

2 a Any two natural links: animals are mostly nocturnal to avoid the heat of the day; insects and animals have adapted to the shortage of food by being active when it rains or by storing food (e.g. camels); plants grow slowly due to the lack of water and in bursts when it rains; plants store water or have long roots to get groundwater; cacti have thick skins to prevent water loss due to the high temperatures.

  b Any one of: people have domesticated animals that can survive in hot deserts such as camels and goats, which then provide food and transport; people were traditionally nomadic, moving from one water source (oasis) to the next so that the scarce water was never used up and they could find food or graze their animals.

3 a Biodiversity is much lower because of the lack of water. Plants and animals find it very difficult to survive, especially in the heat of the daytime, and therefore the food opportunities are reduced – limiting the development of an extensive food web.

  b There is a lack of water, very hot temperatures during the day and even freezing temperatures at night (large diurnal range), these climatic conditions make survival very difficult. Therefore even quite small changes in conditions can lead to the deaths of plants or animals and the destruction of hot desert food webs and alter natural balances.

4 Semi-arid: an area of dry land where the climate is very close to that of a desert, with only a little more rainfall.

Cold environments

1 a Natural links such as: plants grow low to the ground to avoid the wind chill; caribou migrate seasonally to find new pastures and have wide hooves so that they don't sink in marshy areas; plants are small as there are not many nutrients in the soil; the ground is frozen for much

of the year so most plants grow quickly in the brief summer when the top ground layer has thawed. In the food web: bears are omnivores, eating what is available by season, such as berries or fish; polar bears catch seals on the frozen ocean and Arctic foxes catch Arctic hares.

  b One link between nature and people explained, e.g. people have domesticated animals that can survive the cold conditions, such as husky dogs, to pull sleds for transport, or herding of reindeer in Lapland to provide meat, tools and clothing. People also hunt for their food, especially seals and fish such as Arctic char. It is too cold to grow crops so the emphasis is on animals.

2 a Biodiversity is much lower because there is a lack of liquid water, due to lower precipitation and freezing temperatures. Plants and animals therefore find it difficult to survive, especially in winter, and so the food opportunities are reduced – limiting the development of an extensive food web.

  b There is a lack of water, very cold temperatures in winter and at night. These climatic conditions make survival very difficult. Therefore even quite small changes in conditions can lead to the deaths of plants or animals and the destruction of food webs and natural balances. Climate change is a large concern, for example, the lack of sea ice is making it difficult for polar bears to catch their main food source, seals.

3 Answers will vary according to case study. Here the example is Svalbard in the Arctic.

  a There are resources, such as coal, that provide jobs and make money. The fishing grounds provide jobs, food and exports. There is the possibility of developing geothermal energy to provide a cheap source of electricity, hot water and heating. Tourism is another source of income and jobs.

  b Svalbard is a fragile natural environment. It is under a lot of pressure from human activities such as coal mining, which scars the landscape, and pollution from the thermal power station. Overfishing around the islands may disrupt food webs and the ability of fish to repopulate annually. Too much tourism may disturb wildlife or damage the natural landscape.

Physical landscapes in the UK: Review it! (p. 87)

Coastal Landscapes in the UK

1 The three types of mass movement are rock fall, landslide and mudflow.

2 Rock fall is the process where fragments of rock break away from the cliff face, this is often due to processes such as freeze-thaw weathering

3 a A spit and a bar

  b Hurst Castle spit, Slapton Ley is a bar

4 Hard engineering uses artificial structures such as sea walls and groynes to control natural processes, i.e. reduce erosional processes to help protect the coastline.

5 Destructive waves are formed by local storms and are spaced closely together. This results in them being high and they plunge down onto the beach. This creates a strong backwash that removes sand and pebbles and can result in the gradual destruction of the beach.

6 Longshore drift is the movement of sediment along a beach. If waves approach at an angle, sediment will be moved up the beach diagonally and be transported back at right angles to the sea. This will move sediment along the beach in a 'zigzag' pattern.

7 The formation of stacks occur over a long period of time. It begins with a line of weakness in a headland. Energy from the waves erodes the rock along the line of weakness and eventually forms a wave cut notch, which creates a sea cave. Over time, erosion may lead to two back-to-back caves breaking through the headland and forming an arch. The arch will gradually be enlarged through erosion and weathering processes acting on the roof of the arch. The arch wears away and eventually collapses leaving a stack.

River Landscapes in the UK

1 At the source in the upper course, the river is shallow and has a steep sided channel, the valley is V shaped and there is a steep gradient. Waterfalls and interlocking spurs are found here. In the middle course, the valley is U shaped and the channel is wider and deeper. Meanders and floodplains occur in the section. In the lower course, the gradient is more gentle and the valley is very wide and flat. The river is much wider and deeper with a very fast velocity. Here you will find meanders, ox bow lakes, floodplains and levées, leading eventually to the river mouth.

2 Human factors: urbanisation when an increase in impermeable surfaces causes rapid run off; deforestation when a lack of trees means less interception of rainfall and agriculture as water runs more quickly along

ploughed furrows. Physical factors: heavy rainfall causing flash floods; ground becoming saturated after longs periods of rain; impermeable rock that allows for faster surface run off to the river and steep slopes so rain runs down the rivers more quickly.

3   Hydraulic action, where the force of the water hits the river bank and bed. Abrasion, where load carried by the river scrapes along the river bed and bank and wears it away.

4   A gorge is formed where you get a waterfall. The waterfall forms when a river flows over two different rock types. The softer rock, is more easily eroded as it is less resistant. The river erodes the softer rock, creating a plunge pool due to hydraulic action and abrasion. The harder rock is undercut and an overhang is formed, which eventually collapses into the plunge pool. The waterfall retreats upstream as this process is repeated, leaving behind a steep sided gorge, which is very steep sided valley.

Glacial Landscapes in the UK

1   Arête, pyramidal peak and corrie

2   A pyramidal peak is formed when three or more corries are on opposite sides of a mountain top. Each corrie is formed when snow gathers in a sheltered hollow on a mountainside and is compressed into ice. Rotational slip and erosion by abrasion makes the hollow bigger and deeper. A raised lip forms where less erosion takes place at the front of the corrie. The three corries then leave a sharp peak between them called a pyramidal peak.

3   There are often disagreements between developers and conservationists. An example is quarrying of rocks such as limestone and slate. This has economic value, but the quarrying process can lead to pollution of the land and rivers as well as causing a visual eyesore. Tourism can bring conflict between local landowners over access to land. Tourists may affect traffic congestion and cause a rise in house prices.

4   Tourism, farming and quarrying.

5   Advantages: tourists can bring in large amounts of money to an area. In the Lake District in 2014, for example, tourists spent approximately £1000 million in the area. This supported local shops, restaurants and hotels. It also enables the development of new businesses, such as adventure tourism, which can bring jobs and money to the area. Disadvantages: the amount of tourists coming to a glaciated area causes footpath erosion, litter and traffic congestion. Jobs in tourism are often seasonal and very poorly paid.

Urban issues and challenges: Review it! (p. 108)

Global Patterns of urban change

1   A city with a population of more than 10 million people

2   The global population increased slowly up to the 1900s. There was a rapid increase in the 1900s and it doubled over the 30 years between 1970 and 2000 to six billion people. It is now more than seven billion people.

3   People migrate from rural areas to the city due to both push and pull factors. The push factors are: drought, which affects crop yields; sustainable farming not making people enough money and a lack of job opportunities or education services. Pull factors in the developing world, e.g. to Rio de Janeiro, are: the chance of a better quality of life, with better job prospects in factories and better access to education, health care and housing.

Squatter Settlements

1   Social challenges: a lack of sanitation and clean running water, which leads to diseases spreading; a lack of health care and access to doctors means that people who get sick are more likely to die from lack of treatment; housing is very crowded and of poor quality, which also enables the rapid spread of disease.

2   Economic challenges: high levels of employment in the informal sectors, which means low wages; lack of job security and no employee rights; jobs in factories might involve low pay, long hours and sweatshop conditions.

3   Life in a squatter settlement can be improved through self-help schemes, in which communities work together using materials provided by the government to improve the quality of housing. They also have site and service schemes, where the government improves services such as electricity and water supplies to improve the safety and health of the people in the squatter settlement.

Urban Change in the UK

1   'Urban greening' is the process of increasing and preserving open green spaces in cities

2   The advantages of building on brownfield sites are that it regenerates old run-down areas of the city, thus improving the environmental quality of the area, as it reuses land in the inner city for housing or offices.

Brownfield land is cheaper to buy and it prevents greenfield sites from being developed on the edge of the city.

3   London has many opportunities for recreation and entertainment with large numbers of theatres and cinemas for evening entertainment in the West End. There are lots of museums and art galleries for cultural entertainment, as well as many music venues catering to all musical tastes. London has lots of green space and park land such as Hyde Park and Regents Park providing opportunities for recreation, plus lots of sporting facilities such as at the Queen Elizabeth Olympic Park. London is full of different restaurants and pubs for social activities and is very multicultural.

4   Answers will vary according to case study. In London, two contrasting boroughs are Richmond upon Thames in West London and Newham in East London. Richmond has a much higher average income per person; nearly double that of Newham. People in Richmond are more likely to own their own home compared to Newham, where more people rent privately or from the council. Educational achievement in Richmond is higher, with over 65% of students getting 5 A*–C grades at GCSE compared to 50% in Newham.

5   An integrated transport network is important for a city like London, with over 8 million people who need to be able to travel across the city quickly and efficiently. The bus, tube and train networks all link in and use the same ticketing systems and Oyster cards. The new cycle highways also make travel easier with the use of easy access cycle hire from all major stations and numerous other areas.

6   The Lower Lea Valley was chosen because there was lots of cheap, derelict land that could be regenerated. The area had high levels of deprivation and high unemployment so needed an employment boost. The area already had some good rail and tube transport connections.

7   The Olympic regeneration scheme was successful because the development of the new Westfield Centre at Stratford created 10 000 new jobs for local people. The area benefited from the sports facilities and local people are able to use the stadium, aquatic centre and velodrome. Tourism has been boosted in the area with visitors to the Olympic site and to the Queen Elizabeth Park. The former Olympic village used by the athletes has been converted into nearly 3000 homes for local people and includes shops, parks and schools. However, not all local people were happy with the changes as some were forced to move out of the area as the prices had gone up. A long-term plan needs to be in place to ensure local communities don't get forced out of the area as it becomes more wealthy.

How urban change has created challenges

1   Deprivation means a standard of living below that of the majority of people.

2   Three potential challenges for London are: the high levels of unemployment, deprivation and low economic wealth in some areas especially in the east of the city. For example, average wages in West London boroughs like Richmond upon Thames are almost twice that of East London boroughs like Newham. Similarly, life expectancy and education levels are higher in the west than in the east of the city.

3   Environmental challenges: the areas of green belt land being built on due to demands for space. High air pollution levels, which break EU regulations for air quality. A quarter of waste still goes to landfill, which is environmentally polluting as it produces methane.

Urban sustainability in the UK

1   Sustainable means meeting the needs of the city and still being able to provide for future generations.

2   Two inputs are food imported from the countryside or aboard and water piped in from reservoirs. Two outputs are waste sent to landfill outside the city and sewage that goes to sewage treatment works.

3   Answers will vary according to case study. An urban transport scheme such as the sustainable transport network in Bristol has been effective because a range of different strategies were used to encourage people to use alternatives to the car. The public transport system was improved by making the metro rail system to make Bristol more accessible from surrounding towns. The MetroBus provided a rapid method for travelling across the city with reduced journey times. Cycling has been encouraged by having a network of cycle routes to make it much safer and easier to cycle. More 30km/h speed limits make it safer for cyclists and pedestrians. Despite all these measures, people still like the convenience of using their own car when shopping.

4   Settlements can be made sustainable by having affordable housing that is built to be energy efficient with high levels of insulation and triple glazing to keep heat in. By using solar panels and combined heat and power systems, energy usage can be up to 30% less than usual in urban areas. Water can also be saved by recycling rainwater for flushing toilets. Providing sustainable transport options for people such as car sharing, cycle provision and public transport means people will be less likely to use their cars. Travel can also be limited by providing community facilities within the settlement such as schools, health centres and local shops and businesses.

The changing economic world: Review it! (p.140/141)

Economic development and quality of life

1 For example: level of education (e.g. literacy rate), level of health care (e.g. life expectancy, people per doctor), happiness index; political freedom, gender equality; average incomes, access to safe water.

2 **a** and **b**. Some examples are given in Tables 1 (page 111) and 2 (page 115), also see Figure 1 on page 109.

   c LICs and HICs are at different stages of development. LICs have yet to generate enough money that can be spent on making improvements to living conditions or building businesses and industries to perhaps start a multiplier effect. HICs have already been through the growth stage and have wealth to invest in improving people's lives and to help businesses grow.

3 **a** GNI measures the economic development of a country, based on wealth created by businesses and industries inside and outside the country. It is expressed as an average (US$) per person.

   b Infant mortality data can be used to show development by comparing countries; low infant mortality usually means that the living conditions and health care system are of a higher standard than in those countries with a high rate of infant mortality.

   c HDI is a strong measure of development as it uses both social and economic data, and so covers a wider range of development than other single indicators. The data has been collected by the UN for over 25 years, making it a reliable and trustworthy measure. Weaknesses include the lack of data on the condition of the natural environment and human rights.

4 **a** DTM Stage 3: death rate continues to decline, but more slowly, with signs of levelling off. Birth rate declines rapidly from a high level to get nearer to the death rate level.

   b The birth rate in DTM Stage 3 is linked to development changes, such as advances in health care and birth control and the changing role of women. Many women obtain regular jobs and choose to have fewer children and, with more machinery, not as many children are needed to work.

5 **a** A cooler and wetter climate is more suitable for farming, providing enough food for people, sustains fewer diseases and is a more suitable working environment. Places that are too hot, too cold or too dry create difficulties for people.

   b Countries without resources are at a disadvantage in a trading situation, as are countries that are unable to balance their imports and exports. Some trade is unfair: rich countries are more powerful and make more money from trade, while poor countries may lose money when trading.

   c Development aid may cause countries to become dependent on external help so that they do not try to develop themselves, relying too much on the strong external links.

6 **a** Any one country or world region could be chosen, such as India (NEE), which has 22% of its population living on less than US$1.90 a day, or countries such as South Africa (NEE) 16.6%, Haiti (LIC) 54% and Madagascar (LIC) 82%.

   b LICs are too poor to be able to provide a health care system, modern medicines, safe clean water or basic sanitation to all areas. Core urban areas often have better levels of health care, because money is spent in these areas.

   c A cycle of poverty keeps people poor because it is very difficult to break any of the negative factors (e.g. the country lacks money to develop infrastructure), and therefore the parts of the cycle keep influencing each other in a downward spiral. For example, if there are no industries making and selling things then there are no better paid jobs and people stay poor.

7 **a** A refugee is a person who is forced to leave their country because of the strong possibility of death from a natural or human disaster (e.g. drought or war).

   b The push factors out of an LIC include: few better paid jobs, poor health care, lack of schools, disease, and poverty. Pull factors from an HIC include: job opportunities, good health care, better education, security and wealth.

Global development gap

1 **a** Remittances are amounts of money sent to families in home countries by migrants.

   b Infrastructure means the systems, such as roads, power supplies, water supply and sewerage/sanitation networks, etc. These provide the essential framework for industry in an area.

   c Improvements to infrastructure can help industries and businesses establish themselves, employ people and make money. If safe water supplies are available, people are healthier and able to work better. Better education systems help people to have more skills to use in better paid jobs.

   d The parts of the multiplier effect cycle are all positive and therefore have the effect of improving the next part. Each part supports the next throughout the cycle so ensuring that there is continuous improvement.

2 **a** An NGO is an organisation independent of any government, usually a charity, which raises money through donations and has its own aims, such as Oxfam or Save the Children.

   b NGOs usually support smaller development projects and aim to provide what local communities need (a bottom-up approach). Often these use intermediate/appropriate technology; sometimes projects are in the form of emergency aid when there has been a disaster.

   c International aid may not always reduce the development gap when: there is not enough given to make a big difference or there is corruption in the country or community that the aid is given to, so that it doesn't reach the poor people who need it.

3 Two reasons, such as: the level of technology can be understood by people, it is cheaper, it uses less expensive energy, it is easier to repair and it reduces dependency on outside help.

4 Fairtrade may be better as it makes sure that farmers or the people producing goods receive a reasonable price directly. The money they receive is more than if big companies or governments had been involved.

5 **a** Two debt reducing schemes are: 'debt-for-nature' swaps which wipe out some of the debt, and the HIPC initiative (1996), which arranges deferred payments for countries in financial difficulties.

   b Microfinance loans are provided directly to poor people so that they can use the money to improve their lives in the ways that they need to. Loans to a country are usually used for big schemes which often do not directly help poor people in an LIC.

6 Answers will vary according to case study. Here the example is Jamaica.

   a The tourist industry has fluctuated with a dip due to the 2008 world financial crisis, but has recovered. In 2000 1.32m tourists visited the island and this increased to 2.12m in 2015. In 2014 27% of GDP came from tourism and this is predicted to rise to 37.5% in 2025. In 2014 there were 82 500 tourist jobs making up 7.3% of all employment and this is expected to rise to 10.6% in 2025.

   b Benefits: the jobs created and the taxes that the government receives. The wages are then spent in shops and on services, which spreads the wealth around the country, and the government has money to improve the infrastructure such as sewage treatment works.

   c Problems include: reduced amount of tourist money reaching the island because big tourism companies in HICs take a large share. People in areas away from tourist locations still have poor housing and services. The natural environment in areas where tourism is concentrated is damaged by water pollution, unsightly developments and footpath erosion.

Rapid economic development and change

1 **a** Cultural change occurs when there are changes to systems and traditions involving religion, politics, clothing, use of technology, customs and expectations, arts, rules and laws.

   b Economic change is linked to money, wealth, business and industry. Social change is linked to people and their lives, such as quality of life.

2 Answers will vary according to case study. Here the example is Nigeria.

   a Two advantages such as: oil reserves and money made from its export; forest resources such as rubber; farming in wetter areas and grasslands; large young adult workforce; people know English which makes global links easier.

   b Two obstacles such as: widespread corruption in government and business; political conflict between north and south; insurgents; uneven distribution of wealth; ethnic minority groups disadvantaged; poverty; high infant mortality rates; low literacy rate.

3 Answers will vary according to case study. Here the example is Nigeria.

   a One political change, such as: independence from the UK in 1960; recent national elections have been democratic.

   b One environmental change, such as: oil pollution causing damage in the Niger delta area; destruction of tropical rainforests to create space for export crops.

   c One cultural change, such as: the development of Nigerian cinema, known as Nollywood.

   d One social change, such as: slow improvement in quality of life, with HDI increasing from 0.466 in 2005 to 0.514 in 2014.

e Nigeria has experienced both positive and negative changes. Changes include: the improvement in HDI, but its position in the world rank order has not changed. There has been an increase in life expectancy and school attendance. Considerable wealth has been earned from oil exports, but the wealth gap between the north and south of the country has increased. 70% of the population still lives in poverty and there is a high dependence on oil for economic growth (oil is finite and will run out). There have been democratic elections, but the unrest between ethnic groups continues to cause problems. There are environmental issues as a result of development, with deforestation to create agroforestry and farmland and oil pollution.

4 Answers will vary according to case study. Here the example is Nigeria.

a Nigeria is now the 10th largest producer of oil in the world and the oil makes up 91% of exports. In the past, agriculture dominated the economy and most people still work in agriculture. Manufacturing in Nigeria is slowly increasing, with products such as soap, textiles and processed foods.

b TNCs have been both a benefit and a problem for Nigeria. TNCs (such as Unilever) have helped with infrastructure projects, helped develop resources, employed many people and brought money into the country from exports. However, there has been some exploitation of workers (cheap labour), local people were often not considered in decision-making and oil companies have failed to prevent or contain oil spills.

c Nigeria is a Commonwealth country with political links around the world. It trades a lot with other West African nations through ECOWAS and is a member of OPEC as a major oil exporter. There are increasing links with China but trade is strongest with the EU, USA and India.

d Nigeria has oil wealth but this has not spread benefits to all its population. There is a lot of poverty, so aid focuses on social issues such as safe clean water, education and health care.

e Quality of life has not increased rapidly as there has been a lot of corruption; oil wealth has not been invested in social projects or used effectively, environmental damage has made the living conditions of some people (e.g. Ogoni) worse, there has been rapid population growth (doubled since 1990) which has meant that there are more people to provide jobs and services for, and terrorism has emerged in the north of the country (Boko Haram) which has diverted government attention from other issues.

Changes in the UK economy

1 a The quaternary sector consists of research and development and the development and use of information technology.

b Deindustrialisation is the closure of manufacturing industries that have become outdated and unable to compete, e.g. shipbuilding in Newcastle.

2 By the 1980s much of the country's manufacturing industry was old and inefficient. The UK faced competition from countries such as Japan and South Korea that had new factories and the latest technology.

3 Globalisation has helped the UK economy because the UK has a strong service sector, which is a key part of the way in which countries around the world are linked together (e.g. banking), so the UK has a central role. The UK also has developed a strong high-tech industrial sector and so can contribute a lot to the ways in which countries around the world communicate with each other.

4 a Privatisation is where government controlled businesses and industries are sold by the government to become independent.

b FDI can help the UK economy by adding successful businesses and industries from around the world to the country. These provide strong links to other countries, export products and jobs (especially in higher unemployment areas). In this way the country and people gain more money.

c Nationalisation supports industries and businesses experiencing difficulties, ensuring that people keep their jobs, which can be useful in times of recession. However, the government may have to absorb considerable losses of money if the businesses are failing, and it also does not encourage them to become efficient. In comparison, privatisation places decision-making in the hands of the business managers who want to maximise efficiency and profits. In difficult economic times this may mean office and factory closures and the loss of jobs.

5 a IT is helping develop the post-industrial economy because the majority of jobs now involve using a computer or other digital device. There are 1.3m jobs in the IT sector itself. IT gives freedom to businesses to locate anywhere with a broadband connection, which should allow the spread of the post-industrial economy more evenly across the UK.

b Service and finance sectors are helping develop the post-industrial economy by dominating wealth creation: service industries now create 79% of the UKs economic output and this type of employment now

dominates every place in the UK. The finance sector contributes 10% of the UK's GDP each year and there are 2m people employed, especially in key financial centres such as London (which is a world financial centre).

c Research and development is found in science parks because they are located near universities and attract a highly skilled workforce. They are centres for innovation and the development and use of the latest technology. Whilst there is competition between companies, they can share infrastructure, ideas and facilities.

6 Environmental sustainability means looking after the natural environment so that future generations can gain the same benefits from it.

7 Answers will vary according to case study. Here the example is Torr Quarry, Somerset.

a Activities at Torr Quarry have created a large scar in the landscape, and removed natural woodland and grassland, changed run-off patterns and created daily noise and dust.

b The hole in the ground is too big to refill, but the restoration plan aims to achieve some environmental sustainability by creating a lake with some natural features (also used for recreation and water supply). Natural woodlands and grasslands will be replanted. Run-off patterns will not return to what they were, but the lake may help regulate them.

8 a Depopulation is where an area or region is losing people, e.g. the Outer Hebrides. This is generally by migration to other areas due to push-pull factors.

b Retirement migration is where older people, who have retired move to quieter areas such as the seaside or rural locations, e.g. Bournemouth.

9 Answers will vary according to case study. Here the example is South Cambridgeshire.

a It is easy to get to because of the M11 and intercity rail links to London and Cambridge. Stansted international airport is also nearby.

b Four impacts of population growth, such as: more people have brought more congestion to rural roads because people commute to work; communities have changed because the people who move in have an urban lifestyle, rather than the rural farming lifestyle; house prices have increased because demand is higher; higher house prices have pushed young adults away from the area because they cannot afford to buy a home.

10 Answers will vary according to case study. Here the example is the Outer Hebrides.

a There is poor accessibility because the islands are 65 km off the west coast of Scotland and are dependent on ferries. They are a long way from any major city, for example, Glasgow is 470 km away.

b Four impacts of population decline, such as: an ageing population exists because young adults have moved away, as there are few job opportunities or services for them; schools have closed because there are fewer young families, as the young adults have moved away; other services are closing because there are not enough customers to bring in money; farming and fishing industries are in decline because there are fewer people to work in these traditional jobs.

11 Access to peripheral and remoter areas of the UK are being improved by extending fast rail links to the Midlands and North-West (e.g. HS2), and the electrification of northern rail routes. Road improvements to and from peripheral areas, such as upgrading the A303 to the South-West.

12 Improvements to ports (e.g. Felixstowe and Liverpool2) and airport capacities (e.g. Heathrow) are necessary because maximum capacities are about to be reached. If capacity is not increased, then business will be lost to other countries. UK businesses may not be able to import and export as easily, lose money and jobs, and people will have fewer opportunities of travelling abroad.

13 a May vary, areas such as: North-East, North-West, Outer Hebrides, Highlands of Scotland.

b Three differences, such as: lower house prices in the north and higher in the south; lower unemployment in the south and higher in the north; more manufacturing industry in northern regions and more tertiary businesses in southern regions; crofting in Scotland compared with large commercial farms in East Anglia; south is wealthier than the north; northern regions have experienced deindustrialisation and decline while the south has experienced economic growth.

c Two strategies, such as: improvements to transport, e.g. HS2 and Manchester to Sheffield rail link; Enterprise Zones which provide grants and tax incentives to businesses and industries that locate within them in northern areas.

14 a Commonwealth countries are those countries, such as Nigeria, that used to be colonies of the UK but have gained independence, they still maintain strong links with the UK.

b The UK has significant roles in NATO, the UN and the G8 group. (Influence in the EU is likely to decline).

c The UK used to export a lot of manufactured goods but deindustrialisation changed this and so influence as an exporter has declined. The UK continues to be a key importer of goods, being a customer for other countries, so there is some influence here. In contrast, cultural influences have stayed about the same: English is an international language, especially through modern media such as film and many Commonwealth countries still have legal systems and sports based on the UK models.

d The UK is a focal point for world transport and electronic communication because it has major container ports such as Felixstowe, and major airports such as Heathrow, Gatwick and Glasgow. The Channel Tunnel links the UK to mainland Europe. Internet businesses have grown in the UK, partly because of the submarine fibre cables linking it to North America and Europe, and in future to East Asia.

e The UK needs to keep strong political links around the world because the world has been closely interlinked through globalisation. It can gain a lot from links with Commonwealth countries such as support to influence world decisions at the UN, and also help to tackle world issues such as immigration and terrorism.

The challenge of resource management: Review it! (p. 176/177)

Resource management

1 Malnutrition is when people do not have a large enough variety of minerals and vitamins in their diet, which weakens the body and slows development in children.

2 They help people to cook food, heat homes and have light. They also help businesses and industries to operate, providing jobs.

3 a Fertile soils produce crops with higher yields and so make more food available for people.

b Rapid population growth puts a strain on food supplies because there are more people to feed each year increasing demand, meaning there will be less food per person each year if supplies are finite.

4 Food consumption increases because people become wealthier and are able to buy more, and a wider variety of, food.

5 a Mountainous areas are usually wetter because they have relief rainfall, which increases the amount of water available.

b High population densities put pressure on local water supplies, e.g. in London and the south-east of England, as there are lots of people, businesses and industries in a small area all needing water from the same sources.

6 Climate change may reduce water supplies in some parts of the world because it reduces the amount of rainfall. Higher temperatures mean that there is more evaporation, so the amount of water stored in the ground, rivers or lakes (reservoirs) is reduced.

7 a Geothermal energy is created by putting cold water close to magma underground. The heat from the magma turns the water to steam, which is passed through a turbine, generating electricity, or pumped through pipes to provide heating and hot water for buildings.

b Solar energy needs areas with clear skies for sunlight to activate the photovoltaic cells. The heat of the Sun can also be used to heat water in solar panels to provide heating and hot water.

8 World demand for energy has increased because the population has rapidly grown, meaning that more energy is used. More countries are developing economically and have businesses and industries that also use energy.

9 a Agribusiness is a large commercial and intensive farm using a high level of technology with large amounts of money invested in it and owned by a company rather than a family.

b Supermarkets are able to negotiate and buy foods from around the world and get foods that do not grow, or are out of season, in the UK. This enables people to buy more, and a wider variety of, food.

c They are considered to increase the UK's carbon footprint because of the transport emissions by ships, trains or lorries from a foreign country to the supermarket shelves.

10 a UK leisure activities may increase water demand through major uses such as swimming pools and golf courses, and others such as gardening.

b The UK government tries to ensure high water quality through laws to prevent pollution entering the system, for example, from farming or sewage works.

c The UK transfers water from one area to another because the east of the country is drier than the west. However, the major population centres are in the east. The water is transferred from areas of surplus to areas with a deficit.

11 a One reason, such as: oil and natural gas are running out (especially from the North Sea); natural gas supplies are mostly imported and these supplies could get cut off; concern over climate change is reducing the burning of fossil fuels; renewable technologies are being developed rapidly.

b It has declined because fossil fuels have run out or reserves are now low (e.g. only 37% of natural gas reserves left). Also there are now strict international and national regulations on CO_2 emissions.

c Nuclear energy produces radioactivity. Wind turbines may kill wildlife and spoil scenery.

Food

1 a One area, such as: sub-Saharan Africa (including countries such as Sudan, Nigeria, Somalia); Afghanistan, Iraq, Myanmar, North Korea.

b Food grows best where there is sunshine, warmth and plentiful rain, so these areas produce more food than those that are cold or dry.

c They may have problems getting food because there are large numbers of people to share it; people are poor so cannot afford to buy it; food shortages may be caused by climate conditions, pests and diseases, transportation difficulties because of poor infrastructure, lack of water or fighting.

d Conflict disrupts food supplies because farmers are unable to plant, look after or harvest their crops, stores of food may be raided, and transportation may be made very difficult.

2 a Famine is where there is not enough food for the majority of people, which leads to widespread starvation.

b Malnutrition can be considered the biggest killer in the world because it weakens the human body so much that it is unable to fight off illness and disease, so death rates increase.

3 Appropriate technology may be the best way of increasing LIC food supplies because it makes improvements to farming in a way that the people understand and afford.

4 The case study may vary. Here the example is the Indus Basin Irrigation System (IBIS).

a Two features, such as: The Indus Water Treaty (1960) formed an agreement between Pakistan and India; there are large and small dams creating many reservoirs (e.g. Tarbela) with canals to transfer water; there are 14m ha of irrigated farmland.

b Three advantages, such as: increased food supply for Pakistan and India; improved nutrition for Pakistan and India because of the wide range of crops grown; fish farming in the reservoirs provides protein for the diets of people; increased food production has led to food processing industries and export earnings; the dams can also be used to generate HEP.

c Three disadvantages, such as: over-extraction of water in the upper basin limits supplies to the lower basin; evaporation losses from reservoirs and canals is high; salinisation of soils has taken place; waterlogging of fields has taken place.

5 Two features, such as: works with nature; uses only natural fertilisers (e.g. manure); uses a crop rotation system; uses natural food webs to control pests.

6 a Food supplies can be made more sustainable if people in HICs buy only the food they need and eat it before the expiry date. In LIC/NEEs, by storing food correctly and getting food to market more quickly so that it does not rot.

b It is important because catching only controlled quantities and sizes of fish ensures that there are enough fish stocks for the future. Using livestock rearing methods that minimise the damage to soils and forests, keeps the environment in good condition for future generations to use.

7 The case study may vary. Here the example is Jamalpur, Bangladesh.

a Two features, such as: an NGO (Practical Action) helping subsistence farmers; change to a variety of rice that is more resistant to flooding; local fish added to paddy fields; fruits and vegetables grown on the dykes.

b Sustainable food supplies have increased because the new rice variety survives flooding and produces a higher, more reliable yield; fish have added protein to the diet of local people and increased rice yields by adding fertiliser to the soils of the paddy fields; fruit and vegetables have provided people with a more balanced diet.

c The lives of local people have been improved because they are healthier and can therefore work the farms better. A surplus of food is now more likely and this can be sold at the local markets to bring money in that can be spent on improving homes or sending the children to school.

Water

1 a World regions such as the Middle East and Central Asia, and desert and semi-arid countries such as Libya, Saudi Arabia, Morocco, Iran, Afghanistan, Pakistan, Mongolia.

b Wet climates will usually increase the water available while drier climates decrease it. Also high temperatures may increase evaporation and so decrease water availability.

c They may have problems accessing water because the infrastructure is poor. In many areas there is no system for extracting water and getting it to people's homes or for treating the water to make it safe. Poverty means that people have to spend a lot of time collecting and carrying water from sources some distance from their homes.

d People may try to control the supply of water in conflict areas, especially where shared water sources such as rivers pass through more than one region or country. Fighting may mean that water supplies are not available at certain times.

2 a Water stress means that the supplies do not meet the demand, so there is not enough water for all of the users (water deficit).

b Pollution can decrease water supplies because contamination means that the water is unsafe for people to use and they risk illness and death. Industries may not be able to use the water because it affects the quality of their processing.

3 Diverting water from rivers may be best for LICs as the natural flow of the rivers is used to move the water from one place to another so that expensive dams or desalination facilities are not needed. Diversion also avoids displacing people from their homes or flooding farmland with reservoirs.

4 The case study may vary. Here the example is China (south to north).

a Two features, such as: water transfer from lower Yangtze River through existing canals, rivers and lakes; water transfer from central Yangtze River basin to Huang He River; planned to transfer 45 b m³ of water; costs over US$80b.

b Three advantages, such as: reduces water scarcity in the north-eastern areas, e.g. Beijing; supports farms and industries in the north that do not have enough water; helps to keep the economic development of China going.

c Three disadvantages, such as: large scale scheme with a lot of construction (e.g. dams) and a huge cost of over US$80b; there is a lot of water pollution in China and so the water transferred will need to be treated before it can be used; lots of people will be forced to leave their homes to allow construction and flooding by new reservoirs; the water level in the Yangtze will be lower, making it difficult for boats and wildlife.

5 a Two features, such as: collecting rainwater to wash cars or flush toilets; water treated in sewage works can be used for irrigating farmland or for fish farms because harmful bacteria have been removed; reuse of grey water (relatively clean waste water from baths, sinks, etc.)

b Grey water can make supplies more sustainable because it can be collected and reused (e.g. flushing toilets or watering garden). This reduces the demand for clean water and makes supplies last longer.

c Conserving water is important for the future because it will make the supplies of water last longer, especially groundwater. Water infrastructure should be repaired and updated so that losses from leakages are reduced. This will ensure that supplies are increased and maintained in the future for people.

6 The case study may vary. Here the example is Hitosa, Ethiopia.

a Two features, such as: moves water from mountain springs to people in drier area; uses gravity to move the water through a 140 km long pipe; there are 100 stand taps; WaterAid provided half the money; scheme is managed by local communities.

b It has been increased by this scheme as over 65 000 people now have access to 25 l of water per person per day. This is a big improvement on the unreliable seasonal river supplies and one spring.

c Lives have been improved by this scheme because the regular water supplies allow farmers to produce enough food for everyone, so starvation is avoided. People no longer have to walk long distances to collect water, so children can spend more time at school and women can work.

Energy

1 a World regions or countries include: sub-Saharan Africa (such as Sudan, Ethiopia, Mozambique, Namibia), South Asia (such as India and Bangladesh), and Central America (such as Guatemala and Honduras), and Mongolia.

b Geology affects energy availability as processes millions of years ago determined where fossil fuels, uranium and geothermal energy can be found.

c People in an LIC may have poor access to energy because they are too poor to be able to buy it; there are often rapidly growing populations so that there is less energy per person available; the countries are poor and cannot afford to buy energy from other countries or to develop their own resources.

d Political factors may affect energy security because of the decisions made about developing resources within a country, for example, fracking and nuclear energy are controversial; political links with other countries may enable (or not) a country to trade with another to get energy resources; conflict within a country may seriously disrupt energy supplies; environmental laws may change the energy mix of a country.

2 a Energy exports are the sale of energy resources such as oil, natural gas and electricity to other countries, e.g the sale of oil from Saudi Arabia.

b Wind energy needs constant moderate to strong winds and solar needs clear skies for intense sunlight all year. These conditions are not found everywhere.

3 a Environmentally sensitive areas may be threatened because of pressure to produce more energy. There may be exploration and development in sensitive areas, such as oil from the Arctic, or it may lead to more deforestation where poorer people rely on wood for energy, or lead to building more dams and reservoirs which flood natural areas.

b Positives: good for coastal countries; no greenhouse gases are emitted in energy production; electricity is generated twice a day; it is a free resource with little damage to the environment. Negatives: it can be used in limited locations; potential disruption of natural processes; natural areas such as salt-marshes and mudflats may be damaged; large and ugly construction; non-coastal countries or those without tidal estuaries cannot use it.

4 Answers will vary according to case study. Here the example is Amazonia; natural gas, Peru.

a Two features, such as: 385 b m³ of natural gas reserves; financed by foreign companies; pipelines to processing plants (e.g. Cuzco); supplies 95% of Peru's needs and provides exports.

b Three advantages, such as: cheap source of energy, which saves Peruvian people about $1.4b a year; Peruvian government gains revenues from taxes, which it can use to provide services and infrastructure; energy security has been achieved in natural gas and electricity; income from energy sales.

c Three disadvantages, such as: clearance of tropical rainforest for extraction and pipelines; tribal areas have been affected; it is a non-renewable resource; there is foreign involvement. These are disadvantages because the natural environment is damaged and may not recover, the tribal people lose their land and living area and have their way of life disrupted, the natural gas will run out, some profits and benefits are leaving Peru and going to big foreign companies.

5 a Two features, such as: insulating roofs and walls to reduce heat loss and heating does not need to be on or set as high; double or triple glazing to reduce heat loss; energy efficient appliances use less electricity.

b Energy demand can be reduced to achieve sustainability by getting people to use less; laws can make companies produce goods that are more energy efficient, and the use of sustainable transport methods can be developed and people encouraged to use them with incentives.

c Sustainable transport is important because if less energy (e.g. oil) is used, then it will last a lot longer. Reliance on oil can be avoided by replacing car journeys with walking and cycling for short journeys or electric, hybrid or biofuel vehicles for longer journeys. These alternatives diversify the energy mix and make it more sustainable.

6 Answers will vary according to case study. Here the example is Chambamontera, Peru.

a Two features, such as: micro-hydro-electric power scheme; diversion of water from a small river into pipes; gravity used to take water through a turbine; cost was US$45 000; generates 15 kW of electricity (funds from the NGO Practical Action).

b Sustainability has increased because it has reduced the need to cut down trees for firewood so the ecosystem stays intact and soil erosion is prevented. There is little alteration to the river flow so natural processes are unaffected and there is no flooding by a reservoir, there is also no pollution, as it is a renewable energy source.

c There is better health as electricity has replaced the unhealthy kerosene lamps. Schools can now operate computers (and access the internet), which improves education. The health centre can keep medicines longer with refrigeration, and homes and businesses have cheap electricity to use.

Index

The challenge of natural hazards
Tectonic hazards

Study **Figure 1**, a world map showing the distribution of earthquakes and volcanoes.

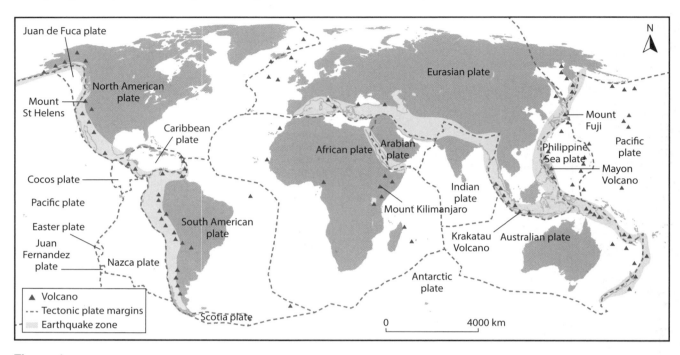

Figure 1

(1) Describe the distribution of earthquake zones shown in **Figure 1**. (2 marks, ★)

....

....

....

....

(2) Outline **one** reason why earthquakes are found in this pattern. (2 marks, ★)

....

....

....

....

NAILIT!

Outline

The command word *outline* means that in your answer you must provide a brief account of relevant information. There is a wide range of geographical factors and processes that could be tested with this type of question, but the answers do not require expansion.

(3) Explain why volcanoes are formed at destructive plate margins. (4 marks, ★★★)

....

....

....

....

....

....

....

....

NAILIT!

Make sure you learn the physical processes happening at each of the three types of plate margin and the landforms that are created. You could draw a flow diagram or a sequence of annotated diagrams in your revision. Include key terms and what they mean.

(4) Explain **two** ways in which people can be protected from an earthquake. (4 marks, ★★)

Protection 1: ....

....

....

....

Protection 2: ....

....

....

....

DOIT!

Reducing the risk of tectonic hazards

Make sure you understand how monitoring, prediction, protection and planning can all reduce the risks of tectonic hazards. Commit to memory how different countries use these four management strategies.

(5) Using a named example, describe the primary and secondary effects of an earthquake. (6 marks, ★★★)

....

....

....

....

....

....

....

....

....

....

....

....

....

WORKIT!

To what extent is it safe to live in an area at risk from a tectonic hazard? (9 marks + 3 SPaG marks, ★★★★★)

'To what extent' questions require you to judge the importance or success of a situation. You need to offer both advantages and disadvantages in your answer.

How to tackle a 'to what extent' question:

Step 1 Give two or three reasons why it may be safe to live in an area at risk from a tectonic hazard, such as Japan.

Step 2 Extend your points by explaining them.

Step 3 Try to include an example to support your point, for example, countries such as Japan have built earthquake-proof buildings.

Step 4 Repeat steps 1–3 for costs of living in an area at risk from a tectonic hazard.

Step 5 Give a conclusion as to whether you think it is safe or not.

> This is a longer written answer and not necessarily straightforward. It may be advisable to make a brief plan for your answer in any available space on the exam paper, to help you link a sequence of points in a fluent way.

NAILIT!

Spelling, punctuation and grammar (SPaG) marks

In the longer answer questions, you will be assessed on your ability to present a fluent answer using correct spelling and punctuation, good-quality grammar (such as sentence structure) and geographical terminology. Make sure that you have prepared for this. Sometimes it can make a difference of a grade!

Weather hazards

Figure 2 shows the aftermath of a tropical storm.

Figure 2

(1) Which **two** statements about tropical storms are true?

Shade **two** circles only. (2 marks, ★)

A Tropical storms develop over the equator. ◯

B Tropical storms form above warm oceans (27°C or above). ◯

C Tropical storms gain energy as they hit land. ◯

D The conditions in the eye of the storm are calm. ◯

E Tropical storms only develop between August and October. ◯

> Remember, answering multiple-choice questions should be quick. Aim to spend no more than 1 minute of exam time to work out the answer to this type of question.

(2) Outline **one** effect of a tropical storm. (2 marks, ★)

....

....

....

....

(3) Describe **two** ways that people can prepare for a tropical storm. (4 marks, ★★)

Preparation 1: ....

....

....

Preparation 2: ....

....

....

DO IT!

Primary and secondary effects

Make sure you understand the difference between primary and secondary effects. Commit to memory the primary and secondary effects of the different natural hazards, such as tropical storms. Learn your case studies!

WORKIT!

Explain how climate change might affect tropical storms. (4 marks, ★★★)

With an 'explain' question, you need to make a point and then give reasons for it. Answers should apply understanding of how climate change can affect a tropical storm – you will not get more than two marks if you only describe how they might change.

How to tackle this question:

Step 1 Describe one way that climate change could affect tropical storms: for example, warmer oceans.

Step 2 Apply understanding by explaining the impact of warmer oceans: for example, warmer oceans may increase the likelihood and intensity of tropical storms.

Step 3 Repeat steps 1–2 with a second point.

(4) For a named tropical storm, assess the extent to which primary effects are more significant than secondary effects.

Use a separate sheet of paper to write your answer to this question. (9 marks + 3 SPaG marks, ★★★★★)

(5) Using a named example or case study, assess the impacts of a recent flood in the UK. (6 marks, ★★★★★)

NAILIT!

Make sure you learn your named examples or case studies so you can provide specific examples to support your answers.

....
....
....
....
....
....
....
....
....
....

(6) 'The UK's weather is becoming more extreme.'

To what extent is this statement true?

Use a separate sheet of paper to write your answer to this question. (9 marks + 3 SPaG marks, ★★★★★)

NAILIT!

To what extent...?

The command phrase in this nine-mark question is *to what extent...?* In your answer, do not just describe and explain – you must also make a judgement on the truth of the situation identified by the question.

Climate change

Study **Figure 3**, a photograph of a volcanic eruption.

Figure 3

(1) Describe **one** way in which volcanic eruptions can affect climate change. (2 marks, ★)

....

....

....

....

(2) State **two** forms of mitigation to manage climate change. (2 marks, ★)

Form 1: ....

....

Form 2: ....

....

> **NAIL**IT!
>
> **State**
>
> The command word *state* is an alternative to 'give' or 'name'. You need to provide a basic response. In some instances, this may be a single word or phrase.

(3) Explain the enhanced greenhouse effect. (4 marks, ★★★)

....

....

....

....

....

....

....

....

....

....

> **NAIL**IT!
>
> A four-mark question is always point marked. This means that you need to give two points and develop them.

(4) Explain the natural causes of climate change. (4 marks, ★★★)

....

....

....

....

....

....

(5) Explain how human activity has contributed to climate change. (6 marks, ★★★)

....

....

....

....

....

....

....

....

....

....

....

NAILIT!

Explain

The command word *explain* is asking you to give reasons for something happening. In Question 5, you need to set out the human causes of climate change and show an understanding of how these activities affect the climate.

WORKIT!

Using named examples, assess the importance of local actions in reducing climate change. (9 marks + 3 SPaG marks, ★★★★★)

An 'assess' question requires you to make an informed judgement. You need to offer **both advantages** and **disadvantages** in your answer.

How to tackle this 'assess' question:

Step 1 Give three ways in which local actions have been effective.

Step 2 Extend your points by explaining how the local actions have been successful.

Step 3 Try to include an example to support each point.

Step 4 Repeat steps 1–3 to explain why global actions are also important in reducing climate change.

Step 5 Add a conclusion to say whether you think local actions can be a success on their own.

The living world
Ecosystems

Study **Figure 1**, which shows a nutrient cycle.

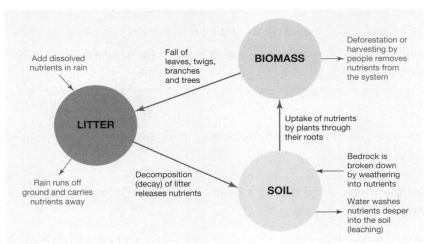

Figure 1

1. Using **Figure 1**, which **one** of the following statements is correct?

   Shade **one** circle only. (1 mark, ★)

   A Nutrients are added to the litter by run-off. ◯

   B Biomass increases as leaves, twigs and branches fall to the ground. ◯

   C Soils get nutrients from the decomposition of litter. ◯

   D Deforestation adds nutrients to the soil. ◯

2. Describe the role of water in the nutrient cycle shown in **Figure 1**. (2 marks, ★)

   ....

   ....

   ....

   ....

WORKIT!

Study **Figure 2**, which shows the Hadley convection cell in the atmosphere.

Describe the weather and climate conditions found at the equatorial location of tropical rainforests shown in **Figure 2**. (2 marks, ★)

How to tackle this question

- When asked to 'describe' something, you should give the main characteristics of it, in this case, of the weather and climate in tropical rainforest regions. No explanation is required.

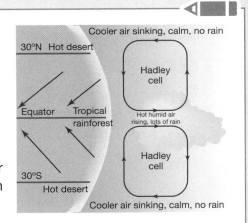

Figure 2

- Study **Figure 2** carefully and use the labels and annotations on the diagram to help you answer the question, along with what you have learned.
- In short-answer questions, the number of marks indicates how many points you should make in your answer. So, two good points here will secure the two marks available.

The climate is hot and humid all year with very little temperature range. Temperatures are usually over 25°C every day, with annual rainfall totals of around 2500 mm.

③ Which **one** of the following statements describes the climate of a tundra area?
Shade **one** circle only. (1 mark, ★)

A Very small temperature range during a year (27°C to 30°C) and high precipitation (approximately 2500 mm a year). ◯

B Small temperature range during a year (20°C to 30°C) and very low precipitation (approximately 125 mm a year). ◯

C Large temperature range during a year (−25°C to +10°C) and low precipitation (approximately 500 mm a year). ◯

D Moderate temperature range during a year (5°C to 18°C) and moderate precipitation (approximately 1500 mm a year). ◯

NAILIT!

Always read questions carefully to make sure that you actually answer the question. For example, do not muddle 'small-scale' with 'large-scale', or 'local ecosystem' with 'global ecosystem'.

④ For a small-scale UK ecosystem you have studied, describe and explain how the natural parts are linked together. (6 marks, ★★★)

Chosen small-scale UK ecosystem: ....

....

....

....

....

....

....

....

....

....

....

DO IT!

Describe and explain

Define what is meant by the command words *describe* and *explain*. If both are included in a question, you must do both to get all of the marks.

(5) For a small-scale UK ecosystem you have studied, describe and explain how the natural balance may be changed. (6 marks, ★★★)

Chosen small-scale UK ecosystem: ....

....

....

....

....

....

....

....

....

....

....

(6) For the global ecosystems you have studied, to what extent do the interactions between the biotic and abiotic parts create a balance within the ecosystems?

Use a separate sheet of paper to write your answer to this question. State your chosen global ecosystems at the start of your answer. (9 marks + 3 SPaG marks, ★★★★★)

NAILIT!

Spelling, punctuation and grammar (SPaG) marks

In the longer answer questions, you will be assessed on your ability to present a fluent answer using correct spelling and punctuation, good-quality grammar (such as sentence structure) and geographical terminology. Make sure that you have prepared for this. Sometimes it can make a difference of a grade!

219

Tropical rainforests

(1) Which **one** of the following statements describes a physical characteristic of a tropical rainforest?

Shade **one** circle only. (1 mark, ★)

A Soils are very fertile with lots of nutrients. ◯

B There are two layers of vegetation in the structure. ◯

C The growing season is eight months a year. ◯

D Nutrient cycling is very fast. ◯

Study **Figure 3**, a photograph of young oil palm trees, planted after deforestation in Malaysia.

Figure 3

(2) For a tropical rainforest you have studied, describe and explain the causes of deforestation. (6 marks, ★★★)

Chosen tropical rainforest: ....

....

....

....

....

....

....

....

....

....

....

....

....

NAILIT!

Make sure that you know the difference between *causes* of deforestation and the *ways* in which deforestation occurs. *Causes* refers to the underlying motives or reasons behind removing trees, rather than the methods or ways in which the trees are removed (such as by logging, road building, HEP construction, farming – although these will be mentioned within the sentences giving the reasons why these human activities are happening). Motives include overcoming poverty and making money for businesses or a country.

Study **Figure 4**, which shows indigenous territories and protected natural areas in countries with tropical rainforest in the Amazon region.

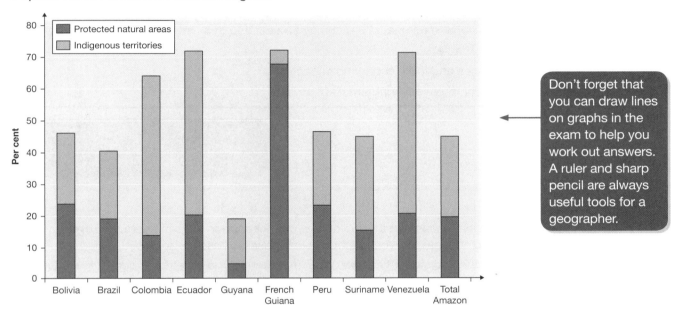

Don't forget that you can draw lines on graphs in the exam to help you work out answers. A ruler and sharp pencil are always useful tools for a geographer.

Figure 4

NAILIT!

Make sure that you know how to read and interpret compound and divided bar graphs. For example, these questions require you to measure only the relevant shaded or coloured section and compare this with the relevant axis scale.

(3) Which country has the largest percentage of protected natural areas? (1 mark, ★)

....

(4) State the number of countries with 30% or more as indigenous territories. (1 mark, ★)

....

(5) Outline **one** possible value to people of the tropical rainforest global-scale ecosystem. (2 marks, ★)

....

....

....

....

6 Outline **one** negative impact of deforestation in an area of tropical rainforest you have studied. (2 marks, ★)

....

....

....

NAILIT!

Many geography questions will ask you about impacts (effects). Remember that impacts can be positive (good) or negative (bad), so always check to see if the question has specified which type of impacts to consider. When thinking about what the impacts are, it may be useful to think about three categories:

- impacts on people (social impacts)
- impacts on money, businesses and jobs (economic impacts)
- impacts on the natural environment (environmental impacts).

> The judgement you make is not right or wrong – it's what you think based on geographical ideas and information.

 DOIT!

Suggest

The command word *suggest* means that there is no one correct answer. Instead, there are several possibilities from which to choose, and you must present a possible case for one.

Study the following question and student answer. What mark would you give the student? How could the answer be improved?

Suggest **one** way in which replanting trees can help make tropical rainforests more sustainable. (2 marks, ★★★)

Replanting trees makes the forest bigger so that there are more plants and animals.

7 Suggest **one** way in which education of people about conservation can help make tropical rainforests more sustainable. (2 marks, ★★★)

....

....

....

....

8 For a tropical rainforest you have studied, explain to what extent this ecosystem provides both opportunities and challenges for development. (9 marks + 3 SPaG marks, ★★★★★)

Use a separate sheet of paper to write your answer to this question. State your chosen tropical rainforest at the start of your answer.

NAILIT!

To what extent...?

The command phrase in this nine-mark question is *to what extent...?* In your answer, do not just describe and explain – you must also make a judgement on the importance or success of the situation identified by the question.

Hot deserts

Answer **either** questions on Hot desert environments **or** questions on Cold environments.

Study **Figure 5**, which shows a climate graph for a hot desert environment.

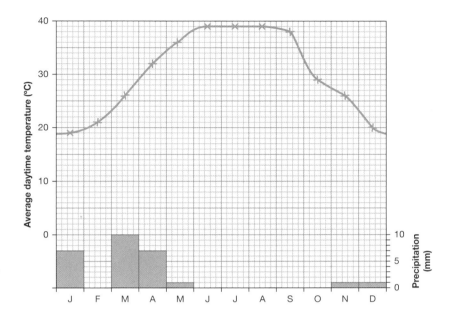

You will have studied *either* Hot desert environments *or* Cold environments. In the exam do **not** answer questions on both.

Figure 5

(1) Using **Figure 5**, which one of the following temperatures is the correct average monthly temperature for October?

Shade **one** circle only. (1 mark, ★)

A 39°C ○

B 32°C ○

C 29°C ○

D 26°C ○

NAILIT!

In questions that require you to write a number as an answer you should **always give the units** for a complete answer. For example, rainfall (or precipitation) is measured in millimetres (mm).

(2) Using **Figure 5**, state the average total rainfall for January. (1 mark, ★)

NAILIT!

Outline

The command word *outline* means that in your answer you must state the main characteristics of something. There is a wide range of geographical factors and processes that could be tested with this type of question, but the answers do not require expansion.

(3) Outline **two** causes of desertification at the edge of some hot deserts. (2 marks, ★)

Cause 1: ....

....

Cause 2: ....

....

(4) Suggest how people may be able to slow or prevent the process of desertification. (4 marks, ★)

....

....

....

....

....

....

....

Study **Figure 6**, a photograph of a hot desert ecosystem.

NAILIT!

Photographs are a useful source of geographical information. Sometimes you may have to look for specific pieces of information in order to answer a question, and sometimes the image may be used as a reminder of what you have learned to help you answer a question (such as **Figure 6** on the left, and Question 5, below).

Figure 6

(5) Using **Figure 6**, explain why a hot desert ecosystem is considered vulnerable. (4 marks, ★★★)

....

....

....

....

....

....

....

(6) Describe and explain the importance of links within a hot desert ecosystem. (6 marks, ★★★★★)

....

....

....

....

....

....

....

....

....

....

(7) For a hot desert environment you have studied, to what extent are the development opportunities for people greater than the development challenges? (9 marks + 3 SPaG marks, ★★★★★)

Use a separate sheet of paper to write your answer to this question. State your chosen environment at the start of your answer.

WORKIT!

For a hot desert environment you have studied, to what extent are the biodiversity issues a result of human or natural pressures? (9 marks + 3 SPaG marks ★★★★★)

How to tackle this question:

- In this question, you will need to include an accurate description, some explanation and a judgement on the relative importance of human and natural pressures.
- Avoid writing lengthy definitions of biodiversity and the pressures – these can be explained in your sentences that answer the question 'to what extent'.

Key ideas to include in an answer to this question:

- The possible changes to biodiversity such as loss of species (plants and animals), extinction of species, changes to habitats and lack of adaptation time.
- Natural pressures, which include the variability in temperature and precipitation.
- For hot deserts, consider the low precipitation (daily and annually) and the high daytime and low night-time temperatures.
- Human pressures include exploitation of resources such as oil, increased population numbers in relatively small areas, modern lifestyle compared to traditional lifestyle, and climate change (enhanced global warming).
- Statements that link biodiversity issues to natural pressures.
- Summary sentences or a conclusion that make a judgement on whether:
  i   human pressures are more responsible for biodiversity issues and by how much, or
  ii natural pressures are more responsible for biodiversity issues and by how much, or
  iii human and natural pressures are equally responsible for biodiversity issues.

Cold environments

Answer **either** questions on Hot desert environments **or** questions on Cold environments.

Study **Figure 7**, which shows a climate graph for a cold environment.

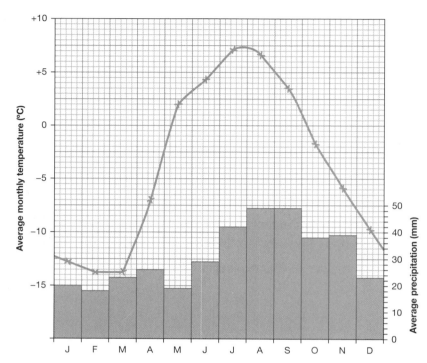

Figure 7

> You will have studied *either* Hot desert environments or Cold environments. In the exam do not answer questions on both!

(1) Using **Figure 7**, which one of the following temperatures is the correct average monthly temperature for December?

Shade **one** circle only. (1 mark, ★)

A +7.1°C   ◯

B −1.7°C   ◯

C −5.9°C   ◯

D −9.9°C   ◯

> **NAILIT!**
>
> Don't forget: in questions that require you to write a number as an answer, you should **always give the units** for a complete answer. For example, rainfall (or precipitation) is measured in millimetres (mm).

(2) Using **Figure 7**, state the average total rainfall for January. (1 mark, ★)

....

(3) Outline how plants and/or animals have adapted to survive in cold environments. (2 marks, ★)

....

....

....

....

....

....

>
>
> **Outline**
>
> The command word *outline* means that in your answer you must state the main characteristics of something. There is a wide range of geographical factors and processes that could be tested with this type of question, but the answers do not require expansion.

Study **Figure 8,** which shows a cold environment ecosystem.

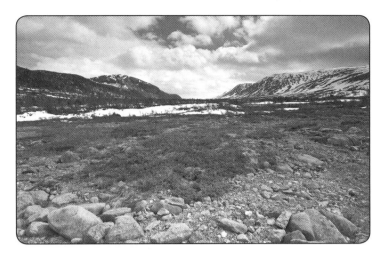

Figure 8

(4) Explain why a cold environment ecosystem is considered vulnerable. (4 marks, ★★★)

....

....

....

....

....

....

....

....

....

(5) Suggest how cold environment wilderness areas can be useful to people. (2 marks, ★)

....

....

....

(6) Describe and explain how problems arising from economic development in cold environments can be reduced. (6 marks, ★★★★★)

....

....

....

....

....

....

....

....

....

....

....

....

(7) For a cold environment you have studied, to what extent are the characteristics of the ecosystem a result of the climatic conditions?

Use a separate sheet of paper to write your answer to this question. State your chosen environment at the start of your answer. (9 marks + 3 SPaG marks, ★★★★★)

WORKIT!

For a cold environment you have studied, to what extent are the biodiversity issues a result of human or natural pressures? (9 marks + 3 SPaG marks ★★★★★)

How to tackle this question:

- In this question, you will need to include an accurate description, some explanation and a judgement on the relative importance of human and natural pressures.
- Avoid writing lengthy definitions of biodiversity and the pressures – these can be explained in your sentences that answer the question 'to what extent'.

Key ideas to include in an answer to this question:

- The possible changes to biodiversity such as loss of species (plants and animals), extinction of species, changes to habitats and lack of adaptation time.
- Natural pressures, which include variability in temperature and precipitation.
- For cold environments, consider wind chill, the low precipitation (daily and annually), and the very low winter temperatures and short summer.
- Human pressures include exploitation of resources such as oil, increased population numbers in relatively small areas, modern lifestyle compared to traditional lifestyle, and climate change (enhanced global warming).
- Statements that link biodiversity issues to natural pressures.
- Summary sentences or a conclusion that make a judgement on whether:
  i   human pressures are more responsible for biodiversity issues and by how much, or
  ii natural pressures are more responsible for biodiversity issues and by how much, or
  iii human and natural pressures are equally responsible for biodiversity issues.

Physical landscapes in the UK
Coastal landscapes in the UK

Answer **two** questions from the following:

questions on Coasts, questions on Rivers, questions on Glacial landscapes.

Shade the circles below to indicate which **two** optional questions you will answer.

> You will have studied *two* topics from Coasts, Rivers and Glacial landscapes. The exam paper will be divided into sections on those three subjects. In the exam, only answer questions on **two** of these topics.

Coasts   Rivers   Glacial ◯

Study **Figure 1**, a 1: 50 000 Ordnance Survey map extract of part of the area around Hurst Castle in Hampshire.

Figure 1

(1) Identify the landform shown in grid squares 3090, 3189 and 3190 in **Figure 1**. (1 mark, ★)

....

(2) Name the process that was involved in creating the landform shown in **Figure 1**. (1 mark, ★)

....

(3) Using **Figure 1**, identify the grid reference for Hurst Castle.

Shade **one** circle only. (1 mark, ★)

A 898318

B 318897

C 908318

D 318908

 NAILIT!

Learn how to read four- and six-figure grid references. A good way of remembering grid references is to always read from the bottom left-hand corner of the map extract and read '*along the corridor and up the stairs*'. (Read horizontal figures followed by vertical figures.)

4 Explain the difference between constructive and destructive waves. (4 marks, ★★★)

....

....

....

....

....

....

....

....

....

....

....

DO IT!

Learn how coastal landforms are created by erosion or deposition. A good way of remembering is by drawing annotated sketches to explain the processes involved in the formation of the landforms. You can also use annotated diagrams to support your answers in the exam. *Annotated* means you label the diagram in detail.

5 Explain the processes involved in the formation of an arch. (4 marks, ★★★)

....

....

....

....

....

....

....

....

6 Explain how landforms are created through deposition along the coast. (6 marks, ★★★★★)

....

....

....

....

....

....

....

....

....

....

....

WORKIT!

Discuss the costs and benefits of using soft engineering to help manage a coastline. (6 marks, ★★★★★)

A 'discuss' question requires you to present **key points** about an idea. You need to offer both advantages and disadvantages in your answer.

How to tackle this 'discuss' question:

Step 1 Give two benefits of soft engineering.

Step 2 Extend your points by explaining how they help to manage a coastline.

Step 3 Try to include a named example or case study to support your point, for example, management strategies used at Wallasea Island, Essex.

Step 4 Repeat steps 1–3 for costs (disadvantages, as well as actual costs) of soft engineering.

Step 5 Draw a conclusion as to whether you think soft engineering is effective in managing a coastline.

River landscapes in the UK

Study **Figure 2**, which shows a storm hydrograph.

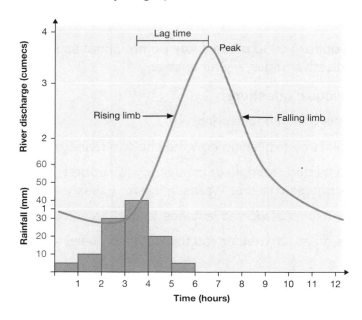

Figure 2

(1) **a** Give a definition of lag time on a hydrograph. (1 mark, ★)

....

....

b Calculate the lag time using the information in **Figure 2**. (1 mark, ★)

....

....

....

....

> Don't forget that you can draw lines on graphs in the exam to help you work out answers. A ruler and sharp pencil are always useful tools for a geographer.

(2) Describe **one** physical and **one** human factor that increase the risk of flooding. (4 marks, ★★★)

....

....

....

....

....

....

....

....

Study **Figure 3**, a 1: 50 000 Ordnance Survey map extract of part of the Wye Valley, Wales.

Figure 3

(3) Using **Figure 3**, give **two** pieces of evidence to show that this is the upper course of a river. (4 marks, ★★)

Evidence 1: ....

....

....

Evidence 2: ....

....

....

NAILIT!

When referring to specific features on an Ordnance Survey map, it is a good idea to give a four- or six-figure grid reference to show where you have found your evidence.

(4) State **two** ways in which the cross section of a river changes from source to mouth. (2 marks, ★)

1 ....

....

2 ....

....

(5) Explain how the size of sediment in a river changes from source to mouth. (4 marks, ★★★)

....

....

....

....

....

....

....

Study **Figure 4**, a photograph of a river at Cuckmere Haven, East Sussex.

Figure 4

(6) Using **Figure 4** and your own knowledge, explain how a floodplain is formed. (4 marks, ★★★)

....
....
....
....
....
....
....
....

NAILIT!

Make sure you learn the formation of all river landforms in four simple steps – this could be done as a flow diagram or a sequence of annotated diagrams in your revision. Include key terms and what they mean.

WORKIT!

Using a named example, assess whether hard engineering has been a success in managing floods in an area of the UK. (6 marks, ★★★★★)

An 'assess' question requires you to make an **informed judgement**. You need to offer both **advantages** and **disadvantages** in your answer.

How to tackle an 'assess' question:

Step 1 Give a definition of hard engineering.

Step 2 Give two ways in which hard engineering has been a success (think economic, social and environmental).

Step 3 Expand each point by explaining how it has helped in managing floods.

Step 4 Try to include an example to support your point, for example, the Jubilee River flood relief channel.

Step 5 Add a conclusion to say whether or not you think it has been a success.

Glacial landscapes in the UK

Study **Figure 5**, a photograph of Striding Edge in the Lake District.

Figure 5

(1) Name the landform shown in **Figure 5**. (1 mark, ★)

....

Study **Figure 6**, a photograph of a glacier.

Figure 6

(2) **a** On **Figure 6**, locate and label these features: (4 marks, ★★★)

Lateral moraine   Terminal moraine   Meltwater   Snout

b State another feature created by glacial erosion (not shown in **Figure 6**). (1 mark, ★)

....

NAILIT!

When labelling a diagram, make sure your arrows are clear and pointing to the exact feature.

(3) Explain the formation of a drumlin. (4 marks ★★★)

....

....

....

....

....

....

....

DOIT!

A diagram would be useful for answering this question. Use your own paper to create one. Make sure you use an arrow to indicate the direction of ice flow.

(4) Describe how economic activities have developed in glaciated upland areas. (4 marks, ★★★)

....

....

....

....

....

....

....

....

....

NAIL**IT**!

A four-mark question is always point marked. This means that you need to give two points and develop them. You should include a named example or case study.

WORK**IT**!

Using a named example, assess whether tourism has been a success in a UK glaciated upland area. (6 marks, ★★★★★)

An 'assess' question requires you to make an **informed judgement**. You need to offer both **advantages** and **disadvantages** in your answer.

How to tackle this 'assess' question:

Step 1 Give two ways in which tourism has been a success (think economic, social and environmental).

Step 2 Extend each point by explaining how it has helped tourism to be successful.

Step 3 Try to include an example to support your point, for example, tourism in the Lake District.

Step 4 Repeat steps 1–3 for the problems created by tourism.

Step 5 Add a conclusion to say whether or not you think it has been a success.

Urban issues and challenges
Global patterns of urban change

Study **Figure 1**, a graph showing world population growth from 1800 to 2050 (predicted).

Human population growth

Figure 1

> Remember, answering multiple-choice questions should be quick. Aim to spend no more than 1 minute of exam time to work out the answer to this type of question.

(1) Using **Figure 1**, which **two** of the following statements about the pattern of population growth are true?

Shade **two** circles only. (2 marks, ★)

A The population of the world has decreased over time. ◯

B The population of the world experienced a rapid growth from 1960. ◯

C Global population is predicted to rise to 9 billion by 2050. ◯

D The peak population growth was reached in 2000. ◯

E The population of the world experienced a rapid growth between 1800 and 1950. ◯

(2) Describe **two** processes that cause a population to increase in size. (4 marks, ★★)

Process 1: ....

....

....

....

Process 2: ....

....

....

....

Study **Figure 2**, a map showing the world's megacities in 2005.

Figure 2

(3) Using **Figure 2**, describe the distribution of the world's largest cities (in 2005). (2 marks, ★)

....

....

....

....

NAILIT!

Describe the distribution

The command phrase in this two-mark question is *describe the distribution*. In your answer, you need to say where the largest cities in the world are.

- Are there more in the northern hemisphere or southern hemisphere?

- Are there more megacities in Asia or elsewhere?

- Are there any continents where there are no megacities?

Study **Figure 3**, a graph showing the urban and rural populations of the world, 1950–2050.

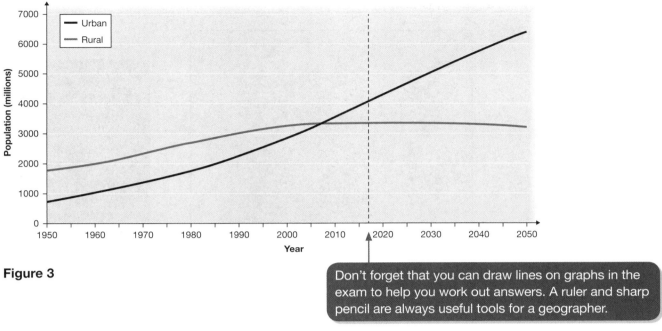

Figure 3

> Don't forget that you can draw lines on graphs in the exam to help you work out answers. A ruler and sharp pencil are always useful tools for a geographer.

(4) Using **Figure 3**, calculate the predicted increase in the urban population from 1950 to 2050. (1 mark, ★)

....

....

(5) Explain why there has been a global increase in urban population and a decline in rural population. (6 marks, ★★★)

....

....

....

....

....

....

....

....

....

....

....

....

Urban growth in LICs and NEEs

Study **Figure 4**, showing a map of Brazil.

Always look carefully at the figures provided in the examination. Some will need to be used directly to extract information and the questions will refer to them, for example, 'Using **Figure 1**'. Others are provided to give you clues and help (as with **Figure 4** in Question 1, below).

Figure 4

(1) Give **two** advantages of Rio de Janeiro's location as a global city. (4 marks, ★★)

Advantage 1: ....

....

....

....

Advantage 2: ....

....

....

....

(2) Explain the problems of working in the informal economy for the people of LIC or NEE cities. (4 marks, ★★)

....

....

....

....

....

....

....

....

Study **Figure 5**, a photograph showing a squatter settlement.

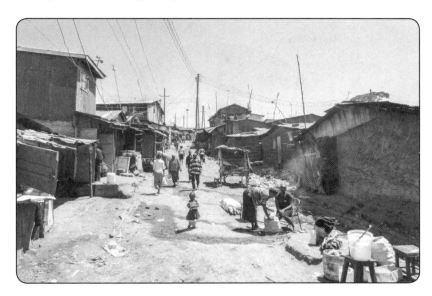

Figure 5

(3) Label the photograph to show the main problems of living in a squatter settlement. (4 marks, ★★)

NAILIT!

Labelling a photograph or diagram

When annotating a photograph, each annotation should give a point and an explanation. For example:

Open sewer, which is unhygienic, may attract rats and this can spread diseases among the people.

WORKIT!

To what extent do squatter settlements in developing countries provide social and economic opportunities for people? (6 marks, ★★★)

This 'to what extent' question is asking how much you agree or disagree with an idea. This means you need to include points **both for** and **against** the idea.

How to tackle this 'to what extent' question:

Step 1 Give **two** social and economic opportunities created by squatter settlements.

Step 2 Extend your points by explaining how they create opportunities.

Step 3 Always include an example to support your point, for example, favelas in Rio de Janeiro.

Step 4 Offer an alternative view or counterargument, for example, why squatter settlements present social and economic challenges for people.

Step 5 Link back to the question as a final summary – do you agree or disagree with the idea that squatter settlements provide social and economic opportunities for people?

Study **Figure 6**, a graph showing methods of transport in Rio de Janeiro.

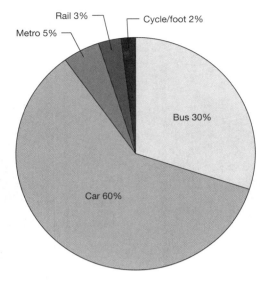

Figure 6

(4) Which transport method is most frequently used in Rio de Janeiro? (1 mark, ★)

....

(5) Describe **two** problems caused by the use of transport in an NEE city like Rio de Janeiro. (4 marks, ★★)

Problem 1: ....

....

....

....

Problem 2: ....

....

....

....

WORKIT!

Evaluate the success of a scheme to improve the quality of life for inhabitants of squatter settlements. Use a named example in your answer. (6 marks, ★★★)

This 'evaluate' question is asking you to say how successful a scheme was in helping to improve the lives of people who live in the squatter settlements.

How to tackle this 'evaluate' question:

Step 1 Give a named example of a scheme, for example, a self-help scheme. It is also useful to include a place where it has been used, such as in the favelas of Rio de Janeiro.

Step 2 Give **two or three** ways in which the scheme works and helps to improve people's lives.

Step 3 Give **two** ways that it hasn't helped people, or state the limitations of the scheme.

Step 4 Give a final overview as a concluding sentence – do you think the successes outweigh the problems?

Urban change in the UK

Study **Figure 7**, a chart showing London's ethnic mix.

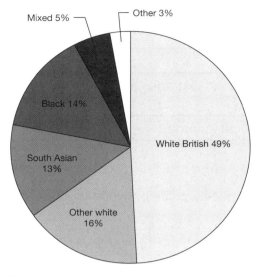

Figure 7

(1) Describe London's ethnic mix as shown in **Figure 7**. (2 marks, ★)

....

....

....

....

NAILIT!

When describing a graph or chart, you should always give the general pattern at the start of your answer and then give data or figures from the graph/chart.

(2) Give **two** reasons why cultural diversity has benefited UK cities. (4 marks, ★★)

Reason 1: ....

....

....

....

....

Reason 2: ....

....

....

....

....

Study **Figure 8**, a table showing the quality of life data for two London boroughs in 2015.

	Richmond upon Thames	Newham
Life expectancy	83 years	77 years
Unemployment	3.8%	8.5%
Pupils achieving 5 A*–C grades at GCSE	70%	57%
Average household income	£46 000	£28 000

Figure 8

(3) Using **Figure 8**, compare the quality of life between the two boroughs. (4 marks, ★★)

....

....

....

....

....

....

....

Study **Figure 9**, a photograph of the Queen Elizabeth Park at the Olympic Park in Stratford.

Figure 9

NAILIT!

Photographs are a useful source of geographical information. Sometimes you may have to look for specific pieces of information in order to answer a question, and sometimes the image may be used as a reminder of what you have learned to help you answer a question (such as **Figure 9** and Question 4, opposite).

④ Using **Figure 9** and an example you have studied, describe how the environment was improved as part of a regeneration project. (4 marks, ★★)

....

....

....

....

....

....

....

....

WORKIT!

Evaluate the effectiveness of an urban regeneration project you have studied. (9 marks + 3 SPaG marks, ★★★★★)

'Evaluate' questions want to know how successful a scheme was, so you need to include points **both for** and **against** the scheme. You could also consider the economic, social and environmental costs and benefits.

SPaG marks give extra marks for your spelling, punctuation and grammar.

How to tackle this 'evaluate' question:

Step 1 Give **three** social, economic and environmental benefits of the scheme.

Step 2 Extend your points by **explaining** the social, economic and environmental benefits.

Step 3 Always include an example to support your points, for example, Newham and the Olympic Park.

Step 4 Repeat steps 1–3, this time giving **three** costs of the scheme (use social, economic and environmental).

Step 5 Give a final conclusion that summarises whether or not the scheme has been a success.

This is a longer written answer and not necessarily straightforward. It may be advisable to make a brief plan for your answer in any available space on the exam paper, to help you link a sequence of points in a fluent way.

NAILIT!

Spelling, punctuation and grammar (SPaG) marks

In the longer answer questions, you will be assessed on your ability to present a fluent answer using correct spelling and punctuation, good-quality grammar (such as sentence structure) and geographical terms. Make sure that you have prepared for this. Sometimes it can make a difference of a grade!

Urban sustainability in the UK

(1) Which of the following is a correct definition for 'urban sustainability'?

Shade **one** circle only. (1 mark, ★)

A Rural areas organised without an over-reliance on urban areas nearby. ◯

B The need for urban areas to consume as much energy and resources as they need to function. ◯

C Cities organised without an over-reliance on the rural area surrounding them, and using renewable energy. ◯

D Urban and rural areas both consume energy and resources to provide for the population living there. ◯

NAILIT!

Multiple-choice questions

While multiple-choice questions are generally regarded as easier questions, it is important not to rush them and make silly mistakes. Look at each option (A, B, C and D) carefully and consider its meaning. You could put a cross over the wording of the incorrect answers as you rule them out – but never put any mark over the circle at the end of the option. Only shade the circle of the answer you have worked out is correct.

---

(2) Identify **two** factors that contribute to a city's ecological footprint. (2 marks, ★)

Factor 1: ....

....

Factor 2: ....

....

(3) Describe **two** features of a sustainable urban community. (4 marks, ★ ★)

Feature 1: ....

....

....

....

Feature 2: ....

....

....

....

....

NAILIT!

A four-mark question is always point marked. This means you need to give two points and develop them.

Study **Figure 10**, a photograph showing a cyclist using the Cycle Superhighway in central London.

Figure 10

(4) Explain how cycling can be an important part of a sustainable transport strategy. (4 marks, ★★)

....

....

....

....

....

....

....

....

....

NAILIT!

Make sure you learn your named examples or case studies so that you can provide specific examples to support your answers.

(5) For a named city you have studied, evaluate the strategies employed to manage traffic congestion. (6 marks, ★★★)

....

....

....

....

....

....

....

....

(6) Give **two** advantages of developing areas of green space in urban areas. (4 marks, ★★)

Advantage 1: ....

....

....

....

Advantage 2: ....

....

....

....

WORKIT!

Suggest ways that cities can be more sustainable. (6 marks, ★★★)

A 'suggest' question requires you to present a possible case for something, in this case, how cities could be made more sustainable.

Step 1 Give **three** ways in which cities can become more sustainable (think social and environmental).

Step 2 Extend your points by explaining how these suggestions are sustainable.

Step 3 Always include an example to support your points, for example, East Village in London.

Step 4 Add a conclusion to say whether or not you think it is possible for cities to be sustainable.

The changing economic world
Economic development and quality of life

Study **Figure 1**, part of a world map showing a classification of countries using average income per person (capita).

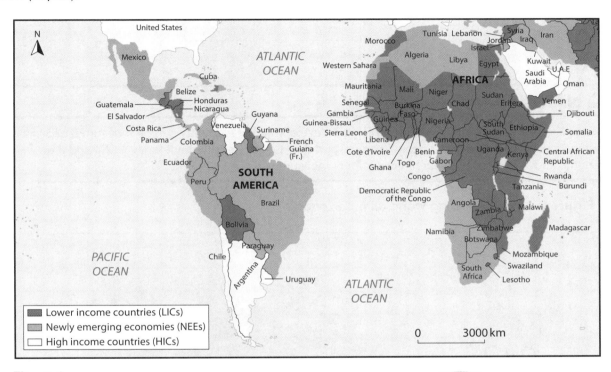

Figure 1

DO IT!

Using keys

Make sure that you always use the keys to maps carefully. In **Figure 1**, for example, in which category is Argentina? In which category is Chad? In which category are Bolivia and Zambia?

(1) Using **Figure 1**, compare the classification of countries in South America with those in Africa. (2 marks, ★)

'Compare' in this question means look for **similarities** and **differences** in the classification categories between South America and Africa. You don't need to count the number of countries in each category. Instead, for example, look for the most common categories, or missing categories.

....

....

....

....

(2) **Figure 1** shows a classification of countries according to income. Outline **one** advantage of using average income per person to measure the development of a country. (2 marks, ★★★)

....

....

....

....

(3) Outline **one** disadvantage of the Human Development Index (HDI) as a measure of a country's development. (2 marks, ★★★)

....

....

....

....

NAILIT!

You need to learn the different ways in which the development of a country can be measured, especially gross national income (GNI) per capita (person) and the Human Development Index (HDI), and the strengths and weaknesses of each development measure.

Study **Figure 2**, a map showing world climate zones.

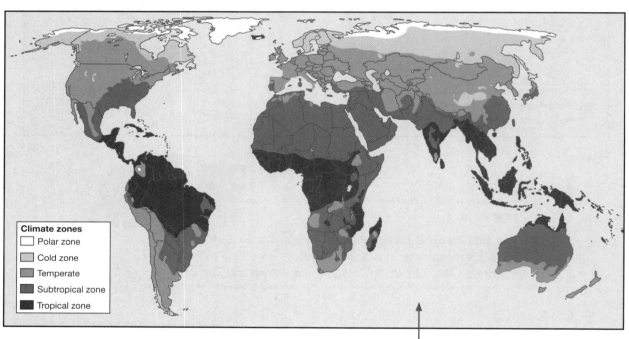

Climate zones
- ☐ Polar zone
- ☐ Cold zone
- ☐ Temperate
- ☐ Subtropical zone
- ☐ Tropical zone

Figure 2

Always look carefully at the figures provided in the exam. Some will need to be used directly to extract information and the questions will usually refer to them, for example, 'Using **Figure 1**' (in Question 1) Others are provided to give you clues and help (as with **Figure 2** in Question 4).

(4) Explain how **one** aspect of physical geography and climate may cause uneven development between countries. (4 marks, ★★★)

....

....

....

....

....

....

....

WORKIT!

Study **Figure 3**, a graph showing the share of world exports (goods) (2005 and 2014) for world regions.

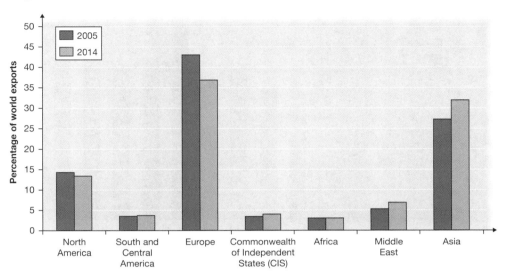

Figure 3

Explain how **one** aspect of economic geography may cause uneven development between countries. (4 marks, ★★★)

How to tackle this question:

The following steps will help you to produce a good standard of answer to this question. You should also study **Figure 3**, which has been included for information.

Step 1 Identify and highlight the command word or phrase. In this case '**explain** how **one aspect**', which means that you must choose one factor linked to economic geography and give reasons for it causing more development in one country than in another.

Step 2 Identify and circle the key ideas in the question. In this case, 'economic geography' and 'uneven development'. Economic geography is to do with making money and uneven development means that some countries get richer while others remain relatively poor.

Step 3 Think of one aspect of economic geography that you have studied in this topic. **Figure 3** has been included to give you a massive clue: **trade**, you could choose the lack of Foreign Direct Investment (FDI), the lack of spending on services such as education and health, or government policies or corruption.

Step 4 In your answer, show how your chosen aspect of economic geography (for example, trade) creates 'richer' countries (HICs and NEEs) and 'poorer' countries (LICs), with two or three well-developed sentences that identify the geographical links.

DOIT!

Give **two** reasons why the suggested Stage 5 of the demographic transition model may be characterised by an ageing population. (2 marks, ★★★)

When asked for two reasons, make sure that they are different and not just the same reason written in a different way!

Study the student answer below, give it a mark out of 2 and suggest how it could be improved:

Stage 5 has a lot of older people because they are living longer. People live longer because there is much better health care.

Study **Figure 4**, a graph showing the number of cases of the disease tuberculosis (per 100 000 people) in 2014 for selected countries.

Note that the graph uses a logarithmic scale.

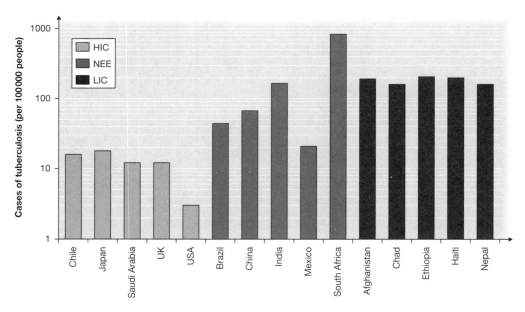

Figure 4

(5) Using **Figure 4**, calculate the difference between the number of cases of tuberculosis (per 100 000 people) in Haiti and the USA. (1 mark, ★)

....

....

(6) Explain why there is such a large variation in the number of cases of tuberculosis between the countries shown in **Figure 4**. (4 marks, ★★★)

....

....

....

....

....

....

....

....

NAILIT!

Sometimes graphs may have a logarithmic (log) scale on one axis (or both axes) when there is a large range in the data to be presented (i.e. very large numbers compared with very small numbers). **Figure 4** has a log scale on the vertical y-axis, showing the number of cases of tuberculosis per 100 000 people in a population. Even though the horizontal lines are spaced differently from a normal graph, each line still represents a unit of measurement. In this case, each line between 1 and 10 increases by 1, each line between 10 and 100 increases by 10, and each line between 100 and 1000 increases by 100. Examiners know that logarithmic graphs are tricky to interpret and will make skills questions based on them as easy as possible.

(7) Suggest **two** reasons why poor health may create a cycle of poverty. (2 marks, ★★★)

Reason 1: ....

....

Reason 2: ....

....

Global development gap

Study **Figure 5**, a graph showing outflowing Foreign Direct Investment (FDI) in $ billions by world group or region 2012–14.

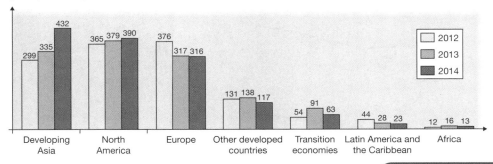

Figure 5

Sometimes there may be several possible answers to a one-mark question, but only one needs to be given. For example, in Question **1a**, change could be simply up or down, or a rate of change, or the size of the change.

(1) a Using **Figure 5**, compare the change in developing Asia with the change in Europe between 2012 and 2014. (1 mark, ★)

....

....

b Calculate the difference between the amount of outflowing FDI from transition economies and the amount outflowing from Africa in 2014. (1 mark, ★)

....

....

(2) Outline **one** way in which FDI can help to increase the development of a country. (2 marks, ★)

....

....

....

....

(3) Suggest **one** reason why FDI is not always successful in reducing poverty within a country. (2 marks, ★★★)

....

....

....

....

NAILIT!

If you are asked for **one** reason, way or problem, and the question is worth two marks, you must provide a fully developed answer. You must explore the **geographical links** within this point in relation to the theme of the question to gain both marks.

Study **Figure 6**, a graph showing the number of tourists visiting Jamaica and the income received from them.

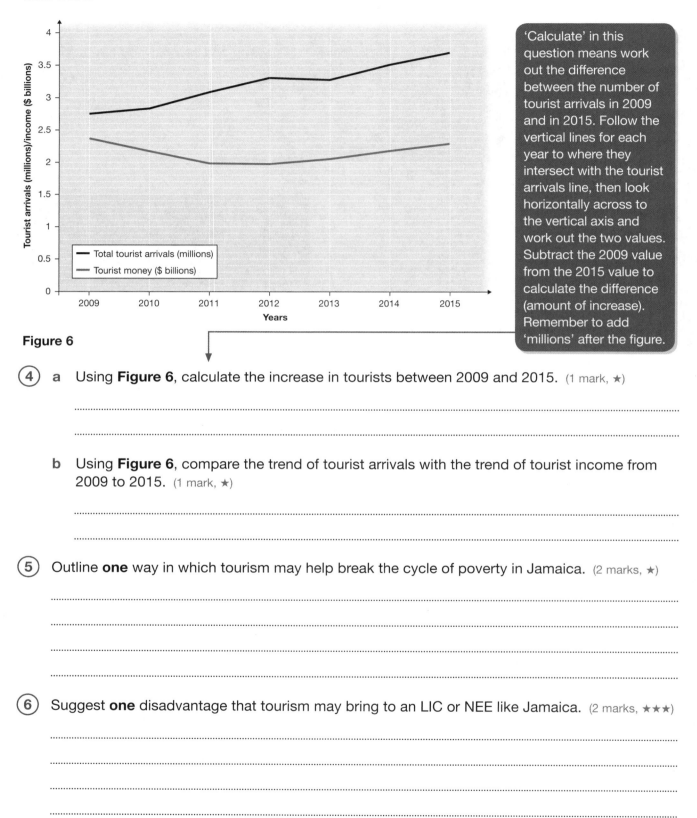

'Calculate' in this question means work out the difference between the number of tourist arrivals in 2009 and in 2015. Follow the vertical lines for each year to where they intersect with the tourist arrivals line, then look horizontally across to the vertical axis and work out the two values. Subtract the 2009 value from the 2015 value to calculate the difference (amount of increase). Remember to add 'millions' after the figure.

Figure 6

(4) **a** Using **Figure 6**, calculate the increase in tourists between 2009 and 2015. (1 mark, ★)

....

....

b Using **Figure 6**, compare the trend of tourist arrivals with the trend of tourist income from 2009 to 2015. (1 mark, ★)

....

....

(5) Outline **one** way in which tourism may help break the cycle of poverty in Jamaica. (2 marks, ★)

....

....

....

....

(6) Suggest **one** disadvantage that tourism may bring to an LIC or NEE like Jamaica. (2 marks, ★★★)

....

....

....

....

NAILIT!

Learn all of the ways in which a country can try to develop economically and improve the quality of life for its people. Think about how successful each of these ways has been or could be in the future.

DOIT!

Justify

Make sure that you know what to do when instructed to 'justify' – this means that you must make a case for your decision. In Question 7, you have a choice between *yes* and *no*, and then you must fully support your answer with as much evidence as possible. There is no right or wrong decision (*yes/no*). Instead, the examiner will be looking at the **quality** of your evidence and your **use** of that evidence to support your decision.

(7) 'Industrial change in an LIC or NEE always brings significant improvements to the quality of life of poor people.'

Do you agree with this statement?   Yes ☐   No ☐

Justify your decision.

Use a separate sheet of paper to write your answer to this question. (9 marks + 3 SPaG marks, ★★★★★)

> This is a longer written answer and not necessarily straightforward. It may be advisable to make a brief plan for your answer in any available space on the exam paper, to help you link a sequence of points in a fluent way.

WORKIT!

'A tourist economy always leads to a range of development benefits for all people in an LIC or NEE.'

Do you agree with this statement?   Yes ☐   No ☐

Justify your decision. (9 marks + 3 SPaG marks, ★★★★★)

How to tackle this question:

- In this question you will need to include some **accurate description**, some **explanation** and a **justification** of your agreement or disagreement with the statement about the scale of benefits to be gained from tourism.
- Avoid just listing all the benefits and problems associated with tourism or just writing everything that you can recall about tourism in the LIC or NEE you have studied. In other words – answer the question.
- Read the statement. Think about what you have learned in this topic and make a decision whether you agree or disagree with it (and shade the relevant box).

Key ideas to include in an answer to this question:

- Recall the ways in which tourism benefits **or** does not benefit all people in an LIC or NEE.
- Recall the country that you have studied where tourism has been used to try to develop the country, including a range of relevant factual material. The case study in the accompanying AQA GCSE Geography Revision Guide covers Jamaica.
- If you think *yes* and agree with the statement, then points you could make include: the range of jobs from unskilled to skilled; the money that this provides for families; the money that this brings into the economy by supporting a range of local businesses (perhaps mentioning the multiplier effect); and to the government in the form of tax money, which is used to improve infrastructure and benefits everyone.
- If you think *no* and disagree with the statement, then points you could make include: that most jobs are low paid and seasonal; income from tourism varies from year to year depending on external influences such as fashion trends and economic conditions; not everyone is able to get a job in tourism and there is a big gap in income and housing quality; and some parts of a country do not benefit at all, as tourism is concentrated in a few areas.

Rapid economic development and change

Study **Figure 7**, a graph showing oil production in Nigeria from the end of 2000 to the start of 2016.

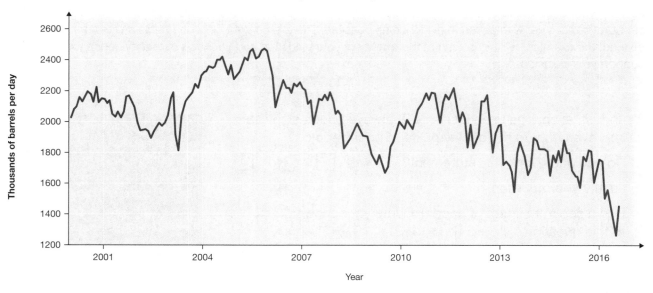

Nigeria's crude oil production

Figure 7

NAILIT!

Make sure that you can read and interpret line graphs by using the axes accurately. Sometimes you will need a ruler and sharp pencil to help work out an accurate answer by drawing lines on the graph to the axes. For example, on the graph above, from the oil production line to the vertical axis showing the amount of oil in thousands of barrels per day. Remember to always state the units of any numerical answers.

① **a** Using **Figure 7**, calculate the decrease in oil production between 2004 and 2016. (1 mark, ★)

....

....

b Using **Figure 7**, compare the trend of oil production between 2004 and 2007 with the trend between 2007 and 2010. (1 mark, ★)

....

....

'Calculate' in this question means work out the difference between the amount of oil production in 2004 and the amount of oil production in 2016. Follow the vertical lines for each year to where they intersect with the production line, then look horizontally across to the vertical axis and work out the two production values. Subtract the 2016 value from the 2004 value to calculate the difference (amount of decrease).

NAILIT!

Remember that if you are asked about 'trends' then there are certain possibilities: either a downward trend (decrease), an upward trend (increase) or no change (stays about the same). However, any of these overall trends may have considerable variations within them – such as lots of 'ups' and 'downs'. There may also be no significant trend with considerable change over time, which can be called a variable trend. Make sure that you stick to the time period(s) given in the question.

(2) For an LIC or NEE country that you have studied, outline **one** political **or** social disadvantage that has restricted positive change in the country. (2 marks, ★)

....

....

....

....

....

....

....

....

NAILIT!

Political, social, cultural and environmental factors

Make sure that you remember examples of each of these geographical factors – identify two for each factor. Then make sure that you use the correct one(s) in an exam question. For example, Question 2 gives you the option of outlining **one** political factor *or* **one** social factor.

---

(3) Explain why the industrial structure of an LIC or NEE country that you have studied has changed. (4 marks, ★★★)

....

....

....

....

....

....

....

....

....

(4) 'International aid and trade are essential to the economic development of LIC or NEE countries.'

Do you agree with this statement?   Yes ☐   No ☐

Justify your decision.

Use a separate sheet of paper to write your answer to this question. (9 marks + 3 SPaG marks, ★★★★★)

NAILIT!

This part of the specification tells you that you must know about a **case study of an LIC or NEE country**. In the accompanying AQA GCSE Geography Revision Guide, this is Nigeria. Make sure that you can recall specific information about Nigeria or the country that you have studied. These longer answer questions do not directly ask for a named country, but you must use your knowledge and understanding of the country you have studied to answer the question, making sure that you include some factual material that you have learned.

WORKIT!

'The quality of life for people in LIC or NEE countries is not always improved by economic development.'

Do you agree with this statement?   Yes ☐   No ☐

Justify your decision. (9 marks + 3 SPaG marks, ★★★★★)

How to tackle this question:

- In this question you will need to include some **accurate description**, some **explanation** and a **justification** of your agreement or disagreement with the statement about the extent to which quality of life is improved by economic development.

- Avoid just listing all the benefits and problems associated with economic development, or writing everything that you can recall about quality of life in the LIC or NEE you have studied. In other words – answer the question.

- Read the statement. Think about what you have learned about this topic and make a decision on whether you agree or disagree with it (and shade the relevant box).

Key ideas to include in an answer to this question:

- Recall the ways in which economic development improves **or** does not improve the quality of life of people in an LIC or NEE.

- Recall the country that you have studied where there has been rapid economic development and change, including a range of relevant factual material. The accompanying AQA GCSE Geography Revision Guide case study covers Nigeria.

- If you think *yes* and agree with the statement, then points you could make include: a slow rate of improvement; low levels of spending on health care and education; poor access to sanitation; rapid population growth increasing pressures; ineffective use of money from trade and resources; civil unrest continuing; and damage to the natural environment.

- If you think *no* and disagree with the statement, then points you could make include: the improvement in quality of life as measured in different ways (such as HDI); longer life expectancy; improvements in access to fresh water and sanitation; higher enrolment in school (boys and girls); and greater income for people.

NAILIT!

Spelling, punctuation and grammar (SPaG) marks

In the longer answer questions, you will be assessed on your ability to present a fluent answer using correct spelling and punctuation, good-quality grammar (such as sentence structure) and geographical terminology. Make sure that you have prepared for this. Sometimes it can make a difference of a grade!

Changes in the UK economy

Study **Figure 8**, a graph showing the trends in employment structure (by industrial sector) in the UK from 1800 to 2000.

Figure 8

① a Using **Figure 8**, compare the change in primary and tertiary employment in the UK between 1800 and 2000. (2 marks, ★)

....

....

....

....

b Give **two** reasons for the decline of secondary industry and secondary employment in the UK towards the end of the industrial stage and into the post-industrial stage (1980 to 2000). (2 marks, ★)

Reason 1: ....

....

Reason 2: ....

....

② Explain how information technology has been creating economic opportunities in the UK since 1990. (4 marks, ★★★)

....

....

....

....

....

....

....

....

DOIT!

③ 'It is not possible to have economic growth in the UK without significant damage to the physical environment.'

Do you agree with this statement?   Yes ☐   No ☐

Justify your decision.

Use a separate sheet of paper to write your answer to this question. (9 marks + 3 SPaG marks, ★★★★★)

> These longer answer questions should be written in a fluent style, showing good use of geographical terminology. Never use lists and bullet points in longer answers.

WORKIT!

'The UK has a very significant role in the wider world.'

Do you agree with this statement?   Yes ☐   No ☐

Justify your decision. (9 marks + 3 SPaG marks, ★★★★★)

How to tackle this question:

- In this question you will need to include some **accurate description**, some **explanation** and a **justification** of your agreement or disagreement with the statement.

- Avoid just listing all the ways that the UK is linked to the rest of the world. Make a judgement on how significant these links are.

- Read the statement. Think about what you have learned in this topic and make a decision on whether you agree or disagree with it (and shade the relevant box).

Key ideas to include in an answer to this question:

- Recall the ways in which the UK has an influence in the rest of the world, perhaps ranking them in order of strength of influence.

- Recall a range of relevant factual material. The accompanying AQA GCSE Geography Revision Guide has a section entitled 'The UK in the wider world'.

- If you think *yes* and agree with the statement, then points you could make include: the UK's position in the UN, NATO and the G8 meetings; the import of goods from all around the world; the colonial legacy of the English language and legal systems; the Commonwealth, immigration links, tourism links; sporting success in the Olympics; internet businesses; and provision of aid to LICs or NEEs.

- If you think *no* and disagree with the statement, then points you could make include: the decline in the UK's role since the days of the British Empire; exit from the EU; controls on immigration; the decline in manufacturing exports; lack of success in some sports; airports reaching full capacity; internet competition; and political and economic competition from NEEs.

NAILIT!

Spelling, punctuation and grammar (SPaG) marks

In the longer answer questions, you will be assessed on your ability to present a fluent answer using correct spelling and punctuation, good-quality grammar (such as sentence structure) and geographical terminology. Make sure that you have prepared for this. Sometimes it can make a difference of a grade!

The challenge of resource management
Resource management

Study **Figure 1**, showing information about world geothermal energy capacity in 2015.

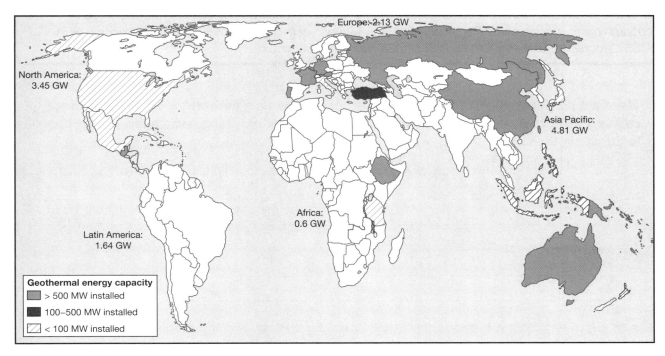

Figure 1

> Make sure that you understand the meaning of the command word *suggest*. It means that there is more than one possible answer. For example, in Question 1 you have to present **one** possible reason.

(1) Suggest **one** reason for the development of geothermal energy in New Zealand. (1 mark, ★)

....

....

(2) Using **Figure 1**, and your geographical knowledge describe the distribution of geothermal energy capacity in the world in 2015. (2 marks, ★)

....

....

....

....

DO IT!

Location and distribution

Make sure that you understand the difference between the geographical terms 'location' and 'distribution'. *Location* is a precise position of something while *distribution* is an amount and spatial arrangement (spread) of something. For distribution, it is best to use compass directions or names of regions (if available).

Study the following student answer to Question 2; give it a mark out of 2 and suggest how it could be improved.

The areas with most geothermal capacity are found on the left of the map, and there are some scattered across the centre of the map.

> ## NAILIT!
>
> Knowledge of geographical terminology, such as *sustainable* and *geothermal energy,* is very important for the exam. This helps you to understand questions, improve the quality of your answers by using correct geographical terms and gain SPaG marks in the questions worth higher marks. The accompanying AQA GCSE Geography Revision Guide contains a Glossary of such terms.

(3) There are proposals to link Iceland to the rest of Europe via a network of undersea electricity cables. Explain why this is considered to be important for providing sustainable energy supplies to Europe in the future. (6 marks, ★★★)

....

....

....

....

....

....

....

....

....

....

....

....

Study **Figure 2**, a graph showing the quality of water in three UK river basins in 2012.

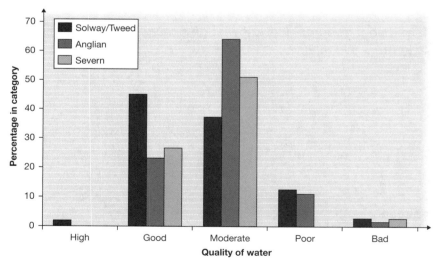

Figure 2

(4) Using **Figure 2**, complete the graph for the River Severn basin using the following data. (1 mark, ★)

Poor   19.7%

⑤ Using **Figure 2**, which **one** of the following statements is correct about the water quality in the three river basins?

Shade **one** circle only. (1 mark, ★)

A The Severn has more moderate water quality than the other two river basins. ◯

B The Anglian has better water quality than the other two river basins. ◯

C The Solway/Tweed has the worst water quality of all three river basins. ◯

D The Solway/Tweed has the best water quality of all three river basins. ◯

Study **Figure 3**, a map showing a possible multi-purpose scheme, proposed in 2013.

Figure 3

⑥ Using **Figure 3**, suggest why water transfer was a major part of this proposed scheme to link the North to the South-East of England. (3 marks, ★★★★★)

....

....

....

....

....

....

Food

Answer **either** questions on Food **or** questions on Water **or** questions on Energy.

Shade the circle below to indicate which optional questions you will answer.

You will have studied *either* Food *or* Water *or* Energy. The exam paper will be divided into sections on Food, Water and Energy. In the exam, only answer questions on **one** of these topics.

Shade **one** circle only

Food ⦿   Water ⦿   Energy ⦿

Study **Figure 4**, a map showing daily calorie intake per capita in the world in 2015.

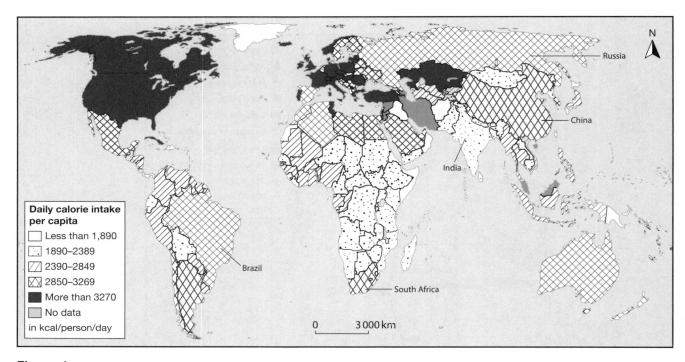

Daily calorie intake per capita
- ☐ Less than 1,890
- 1890–2389
- 2390–2849
- 2850–3269
- More than 3270
- No data

in kcal/person/day

Figure 4

(1) What was the range of daily calorie intake per capita in India?

Shade **one** circle only. (1 mark, ★)

A Less than 1890 kilocalories ⦿   **C** 2390–2849 kilocalories ⦿

B 1890–2389 kilocalories ⦿   **D** 2850–3269 kilocalories ⦿

(2) Using **Figure 4**, describe the distribution of daily calorie intake per capita in South America. (2 marks, ★)

....

....

....

....

(3) Give **two** natural (physical geography) factors that may affect food supply in a country. (2 marks, ★)

Natural factor 1: ....

....

Natural factor 2: ....

....

> Make sure that you do not muddle 'small-scale schemes' with 'large-scale schemes'. Also, make sure you know what is meant by the term 'sustainability'.

WORKIT!

Explain how small-scale schemes can create sustainable food supplies. (6 marks, ★★★)

How to tackle this question:

- You will need to recall the example of a sustainable small-scale scheme that you have studied. In the accompanying AQA GCSE Geography Revision Guide, the case study is Jamalpur in Bangladesh.
- You should also be able to bring into your answer other relevant ways of achieving sustainable food supplies through small-scale schemes.
- You should make clear links between the features of the small-scale schemes and how they bring about sustainability.
- Sustainability points may include: ensuring that food supplies exist throughout a year and from year to year; reducing hunger and malnutrition; minimising damage to soils and reducing water use; and ensuring some income for poor farmers.

(4) Give **two** reasons why food consumption is increasing in the world. (2 marks, ★)

Reason 1: ....

....

Reason 2: ....

....

(5) Explain how the impacts of food insecurity cause difficulties for poorer people. (6 marks, ★★★)

....

....

....

....

....

....

....

....

NAILIT!

- Remember to identify the **command word** or **phrase** in a question. In Question 5, this is *Explain*. 'Explain' means to set out the reasons.
- Identify the **key geographical ideas** in a question. In Question 5, these are *impacts of food insecurity* and *poorer people*.
- Circle both the command words or phrases and the key geographical ideas in questions to help keep you on track.

Water

Study **Figure 5**, a map showing water stress in the densely populated river basins of South and East Asia.

Figure 5

1. What was the water stress level in the Indus River basin?

   Shade **one** circle only. (1 mark, ★)

   A Low to medium stress ◯

   B Medium to high stress ◯

   C High stress ◯

   D Extremely high stress ◯

   > Remember, answering multiple-choice questions should be quick. Aim to spend no more than one minute of exam time to work out the answer to this type of question.

2. Using **Figure 5**, describe the distribution of water stress in river basins in China. (2 marks, ★)

   ....

   ....

   ....

   ....

3. Give **two** reasons why water consumption is increasing in the world. (2 marks, ★)

   Reason 1: ....

   ....

   Reason 2: ....

   ....

WORKIT!

Explain how small-scale schemes can create sustainable water supplies. (6 marks, ★★★)

How to tackle this question:

- You will need to recall the example of a sustainable small-scale scheme that you have studied. In the accompanying AQA GCSE Geography Revision Guide, the case study is Hitosa in Ethiopia.
- You should also be able to bring into your answer other relevant ways of achieving sustainable water supplies through small-scale schemes.
- You should make clear links between the features of the small-scale schemes and how they bring about sustainability.
- Sustainability points may include: ensuring reliable water supplies throughout a year and from year to year; maximising water storage and minimising water wastage; ensuring that water is clean so that people are healthier; ensuring that clean water is shared fairly between all people; and keeping costs of water supply low.

(4) Give **two** reasons why groundwater management helps to create sustainable water supplies. (2 marks, ★)

Reason 1: ....

....

Reason 2: ....

....

NAILIT!

- Remember to identify the **command word** or **phrase** in a question. In Question 5, this is *Explain*. 'Explain' means to set out the reasons.
- Identify the **key geographical ideas** in a question. In Question 5, these are *impacts of water insecurity* and *poorer people*.
- Circle both the command words or phrases and the key geographical ideas in questions to help keep you on track.

(5) Explain how the impacts of water insecurity cause difficulties for poorer people. (6 marks, ★★★)

....

....

....

....

....

....

....

....

Energy

Study **Figure 6**, a map showing energy access and security for countries in 2015.

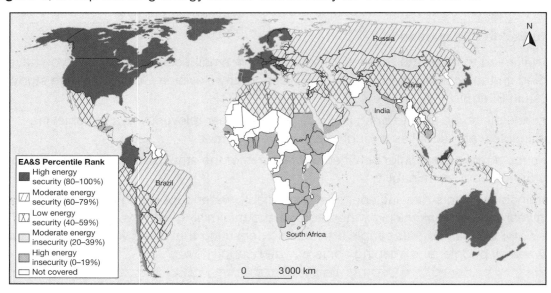

Figure 6

(1) What was the energy access and security performance level of China in 2015?

Shade **one** circle only. (1 mark, ★)

A 0–19% ◯

B 20–39% ◯

C 40–59% ◯

D 60–79% ◯

(2) Using **Figure 6**, describe the distribution of energy access and security performance levels in South America in 2015. (2 marks, ★)

....

....

....

....

(3) Give **two** reasons why home design can help to create sustainable energy supplies. (2 marks, ★)

Reason 1: ....

....

Reason 2: ....

....

NAILIT!

- Remember to identify the **command word** or **phrase** in a question. In Question 4, this is *Explain*. 'Explain' means to set out the reasons.
- Identify the **key geographical ideas** in a question. In Question 4, these are *impacts of energy insecurity* and *poorer people*.
- Circle both the command words or phrases and the key geographical ideas in questions to help keep you on track.

(4) Explain how the impacts of energy insecurity cause difficulties for poorer people. (6 marks, ★★★)

....

....

....

....

....

....

....

....

....

....

....

....

(5) Give **two** reasons why energy consumption is increasing in the world. (2 marks, ★)

Reason 1: ....

....

Reason 2: ....

....

> Make sure that you do not muddle 'small-scale schemes' with 'large-scale schemes'. Also, make sure you know what is meant by the term 'sustainability'.

WORKIT!

Explain how small-scale renewable energy schemes can create sustainable energy supplies. (6 marks, ★★★)

How to tackle this question:

- You will need to recall the example of a sustainable small-scale scheme that you have studied. In the accompanying AQA GCSE Geography Revision Guide, the case study is Chambamontera in Peru.
- You should also be able to bring into your answer other relevant ways of achieving sustainable energy supplies through small-scale schemes.
- You should make clear links between the features of the small-scale schemes and how they bring about sustainability.
- Sustainability points may include: ensuring a reliable energy supply; low set-up and maintenance costs for local people; low energy costs for local people; use of renewable energy sources that minimise negative impacts on the local and wider environment; and improvements in health levels through lower pollution.

Paper 1 Section A: The challenge of natural hazards

Answer all questions in this section.

Question 1 The challenge of natural hazards

Study **Figure 1**, a graph showing changes in sea surface temperature and hurricane strength.

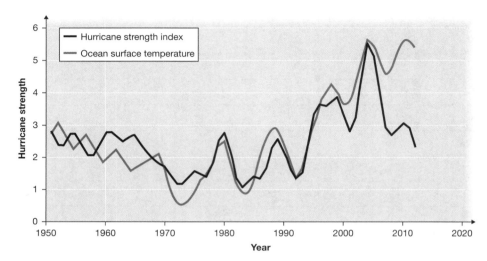

Figure 1

1.1   Describe **one** change in the sea surface temperature shown in Figure 1.

....

....

....

....

[2 marks]

1.2   Outline **one** reason why sea temperatures have changed over time.

....

....

....

....

[2 marks]

Study **Figure 2**, a graph showing changes in global temperature, 1880–2014.

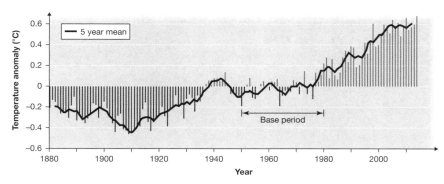

Figure 2

1.3   Using **Figure 2**, which **two** statements are true?

Shade **two** circles only.

A Global temperature has increased since 1880.   ◯

B Global temperature was lower in 2000 than it was in 1880.   ◯

C At its maximum, the global temperature was approximately 0.6°C
   above the base temperature.   ◯

D At its minimum, the global temperature was approximately 0.6°C
   below the base temperature.   ◯

E The global temperature starts to decrease from 1980.   ◯

[2 marks]

1.4   Explain the natural causes of climate change.

....

....

....

....

....

....

....

....

....

....

....

....

[6 marks]

Study **Figure 3**, a map showing the track of Hurricane Katrina in 2005.

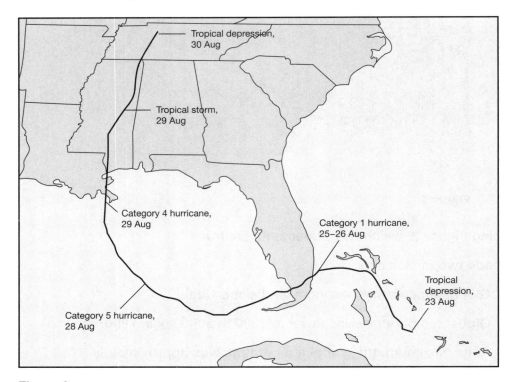

Figure 3

1.5   Describe the direction of the track of Hurricane Katrina in **Figure 3**.

....

....

....

....

[2 marks]

1.6   Describe **one** feature of a tropical storm.

....

....

....

....

[2 marks]

Study **Figure 4**, a photograph showing an area affected by Typhoon Haiyan.

Figure 4

1.7   Using **Figure 4**, describe the primary and secondary effects of a tropical storm.

....

....

....

....

....

....

....

....

[4 marks]

1.8   State **two** ways in which climate change might affect tropical storms.

Effect 1: ....

....

Effect 2: ....

....

[2 marks]

1.9   Choose **either** a volcanic eruption **or** an earthquake.

Evaluate the success of responses to this tectonic hazard in two areas with contrasting levels of wealth.

[9 marks + 3 SPaG marks]

Use a separate sheet of paper to write your answer to this question. State your chosen tectonic hazard at the start of your answer.

Paper 1 Section B: The living world

Answer all questions in this section.

Question 2 The living world

Study **Figure 5**, which shows part of a tropical rainforest food web.

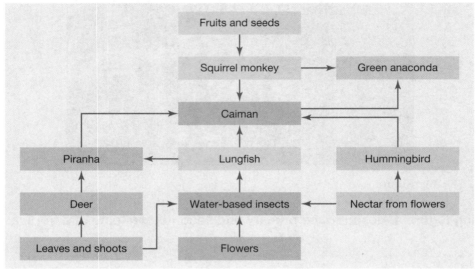

Figure 5

2.1   Using **Figure 5**, which one of the following statements is correct?
Shade **one** circle only.

A The deer eat the piranha.

B The green anaconda eats squirrel monkeys and caimans.

C The caimans eat green anacondas and hummingbirds.

D The water-based insects get their food from the lungfish.

[1 mark]

2.2   Describe what would happen in the food web shown in **Figure 5** if the caiman became extinct.

....

....

....

....

[2 marks]

2.3   Which **one** of the following statements correctly describes the characteristics of a hot desert?
Shade **one** circle only.

A Large average temperature range during a year (−25°C to +10°C) and low precipitation (approximately 500 mm a year).

B Very small average temperature range during a year (27°C to 30°C) and high precipitation (approximately 2500 mm a year).

C Small average temperature range during a year (20°C to 30°C) and very low precipitation (approximately 125mm a year).

D Moderate average temperature range during a year (5°C to 18°C) and moderate precipitation (approximately 1500mm a year).

[1 mark]

Study **Figure 6**, a diagram showing the structure of a tropical rainforest in northern Australia.

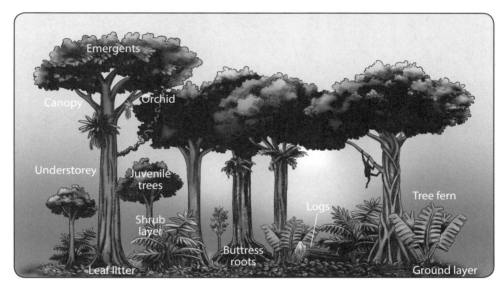

Figure 6

2.4   Describe and explain the structure of the tropical rainforest shown in **Figure 6**.

....

....

....

....

....

....

....

....

....

....

....

....

[6 marks]

Study **Figure 7**, a graph showing deforestation in the Amazon tropical rainforest, Brazil, from 1988 to 2014.

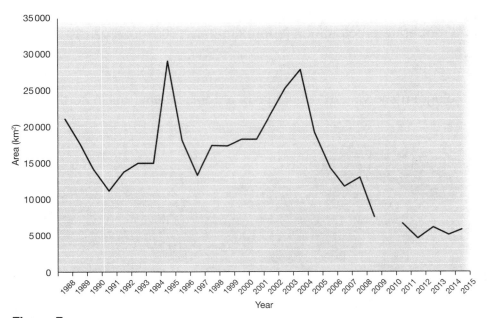

Figure 7

2.5   In 2010, the deforestation rate was 7000 km². Plot this on **Figure 7** and complete the line.   **[1 mark]**

2.6   In which year was the highest rate of deforestation?

....   **[1 mark]**

2.7   Outline **one** impact of deforestation on the economic development of a country you have studied with a tropical rainforest ecosystem.

....

....

....

....   **[2 marks]**

2.8   Suggest **one** way in which debt reduction can help make tropical rainforests more sustainable.

....

....

....

....   **[2 marks]**

2.9   For a hot desert environment **or** cold environment you have studied, to what extent has technology helped balance economic development and conservation in the environment?   **[9 marks]**

Use a separate sheet of paper to write your answer to this question. State your chosen environment at the start of your answer.

Paper 1 Section C: Physical landscapes in the UK

Answer **two** questions from the following:

Question 3 (Coasts), Question 4 (Rivers), Question 5 (Glacial)

Question 3: Coastal landscapes in the UK

3.1   Describe **one** process of coastal erosion.

....

....

....

....

                                                                                                        [2 marks]

Study **Figure 8**, a photograph of Old Harry Rocks, Dorset.

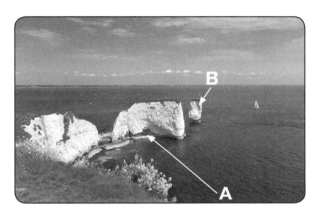

Figure 8

3.2   Using **Figure 8**, identify the coastal features shown at **A** and **B**.

           Feature **A**: ....

           Feature **B**: ....

                                                                                                        [2 marks]

3.3   Using **Figure 8**, explain how coastal feature **B** was formed.

....

....

....

....

....

....

....

                                                                                                        [4 marks] 277

Study **Figure 9**, a photograph of coastal defences at Scarborough.

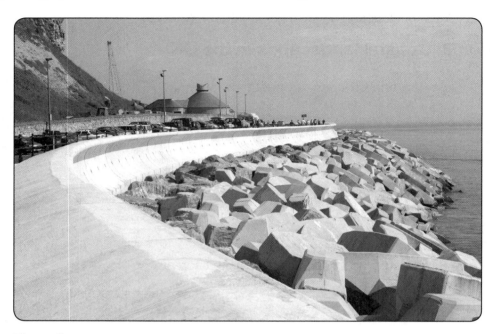

Figure 9

3.4   Name **one** coastal defence shown in **Figure 9**.

....

[1 mark]

3.5   Using **Figure 9** and your own knowledge, explain the advantages and disadvantages
           of hard engineering to protect the coastline.

....

....

....

....

....

....

....

....

....

....

....

[6 marks]

Question 4: River landscapes in the UK

Study **Figure 10**, showing a section of a river.

4.1   What stage of the river's long profile is shown in **Figure 10**?

....

....

....

....

                                                                [1 mark]

Figure 10

Study **Figure 11**, a photograph of a meandering river.

Figure 11

4.2   Explain the formation of the landform shown in **Figure 11**.

....

....

....

....

....

....

....

....

                                                                                        [4 marks]

4.3   Describe **two** causes of river flooding.

Cause 1: ....

....

Cause 2: ....

....

[4 marks]

4.4   Using a named example, explain the costs and benefits of a flood management scheme in the UK.

....

....

....

....

....

....

....

....

....

....

....

....

[6 marks]

Question 5: Glacial landscapes in the UK

Study **Figure 12**, a photograph showing a glacial landform.

5.1   Name the glacial landform shown at **A** in **Figure 12**.

....

....

....

....

[1 mark]

Figure 12

Study **Figure 13**, a 1: 50 000 Ordnance Survey map extract of part of the Lake District.

Figure 13

5.2   Identify the glacial feature at grid reference 345151. Shade **one** circle only:

A truncated spur   ⃝   **C** arête   ⃝

B glacial trough   ⃝   **D** moraine   ⃝

[1 mark]

Study **Figure 14**, a photograph of the glacial lake in grid square 3415 in **Figure 13**.

Figure 14

5.3   Using **Figures 13** and **14,** name the lake shown in the photograph.

....

[1 mark]

5.4   Describe **two** ways in which glaciated areas can create economic opportunities.

....

....

[2 marks]

5.5   Suggest why you may get conflicts when developing a glacial upland area.

....

....

....

....

....

....

....

[4 marks]

5.6   Explain how different erosional landforms may be created in a glacial environment.

....

....

....

....

....

....

....

....

....

....

....

[6 marks]

Paper 2 Section A: Urban issues and challenges

Answer all questions in this section.

Question 1: Urban issues and challenges

Study **Figure 1**, a map from 2005 showing the world's largest megacities.

Figure 1

1.1   Using **Figure 1**, which **two** of the following statements are true?

Shade **two** circles only.

A Most of the world's largest megacities are located in HICs.   ○

B Most of the world's largest megacities are located in Asia.   ○

C Most of the world's megacities are located in LICs or NEEs.   ○

D Europe has the megacities with the largest population in the world.   ○

E Africa does not have any megacities.   ○

[2 marks]

Study **Figure 2**, a map showing the percentage of urban population in different parts of the world.

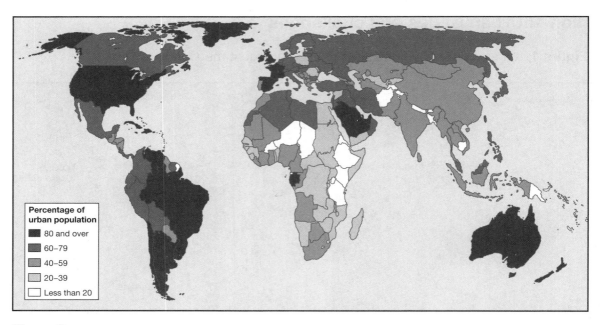

Figure 2

1.2   Using **Figure 2**, describe the global pattern of urban population.

....

....

....

....

[2 marks]

1.3   Suggest **two** reasons why the rate of urban growth is increasing more rapidly in developing countries.

Reason 1: ....

....

Reason 2: ....

....

[2 marks]

1.4   Discuss the social and economic effects of rapid population growth on a city in a lower income country (LIC) or newly emerging economy (NEE) you have studied.

....

....

....

....

....

....

....

....

....

....

....

[6 marks]

Study **Figure 3**, a table comparing two London boroughs in 2015.

	Richmond upon Thames	**Newham**
Life expectancy	83 years	77 years
Unemployment	3.8%	8.5%
Pupils achieving 5 A*–C grades at GCSE	70%	57%
Average household income	£46 000	£28 000

Figure 3

1.5   Using **Figure 3**, describe **two** differences in levels of deprivation between Richmond upon Thames and Newham.

Difference 1: ....

....

....

....

Difference 2: ....

....

....

....

[4 marks]

1.6   State **two** ways in which traffic can be managed in an urban area.

1: ....

....

2: ....

....

[2 marks]

1.7   Using a named example, explain how a city can reduce its use of resources to improve its sustainability.

....

....

....

....

....

....

....

....

....

....

....

[6 marks]

1.8   Evaluate the success of a UK regeneration project you have studied.

[9 marks + 3 SPaG marks]

Use a separate sheet of paper to write your answer to this question.

Paper 2 Section B: The changing economic world

Answer all questions in this section.

Question 2: The changing economic world

Study **Figure 4**, a world map showing life expectancy in countries of the world in 2015.

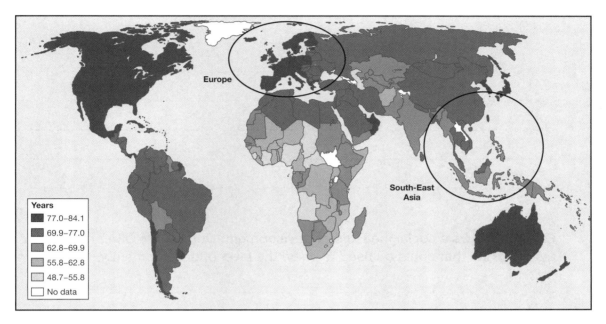

Figure 4

2.1   Using **Figure 4**, compare the life expectancies in Europe with those in South-East Asia.

....

....

....

....

[2 marks]

Study **Figure 5**, a scatter graph showing GNI per capita (person) and infant mortality rates for selected countries in 2015.

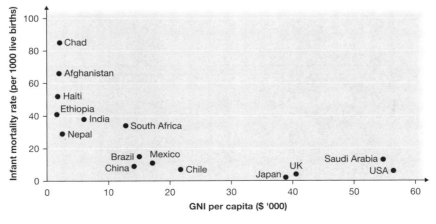

Figure 5

2.2   Using **Figure 5,** explain how infant mortality rates can be used to show differences in
         quality of life between countries.

....

....

....

....

....

....

....

[4 marks]

2.3   **Figure 5** shows a social measure of development. Outline **one** other measure of
         quality of life that could be used to show the level of development within a country.

....

....

....

....

[2 marks]

Study **Figure 6**, maps showing the migration of people from the UK and Nigeria to other continents and
countries.

Figure 6

2.4   Using **Figure 6**, name **one** continent that is receiving people from the UK.

....

[1 mark]

2.5   Using **Figure 6**, explain why the patterns of migration out of the UK and out of Nigeria are different.

....

....

....

....

....

....

....

....

[4 marks]

2.6   Give **two** reasons why there is a disparity in the wealth of people within countries.

Reason 1: ....

....

Reason 2: ....

....

[2 marks]

Study **Figure 7**, a line graph showing changes in the UK's industrial employment.

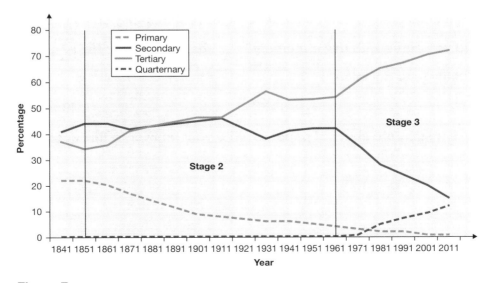

Figure 7

2.7   Compare the changes to primary and tertiary employment during Stage 2 in **Figure 7**.

....

....

[1 mark]

2.8   Using **Figure 7**, calculate the percentage difference between primary and quaternary employment in 2011.

....

....

[1 mark]

2.9   Outline **one** way in which globalisation has helped to increase tertiary employment and business in the UK.

....

....

....

....

[2 marks]

2.10   Suggest **one** problem that deindustrialisation has caused in the UK.

....

....

....

....

[2 marks]

2.11   'Industrial activity can be made to be environmentally sustainable in the UK.'

Do you agree with this statement?   Yes ☐   No ☐

Justify your decision.

[9 marks + 3 SPaG marks]

Use a separate sheet of paper to write your answer to this question.

Paper 2 Section C: The challenge of resource management

Answer Question 3 and either Question 4 or Question 5 or Question 6.

Question 3: The challenge of resource management

Study **Figure 8**, which provides information about undernourishment and malnutrition in the world.

3.1 What has been the decline in undernourishment since 1992?

....

....

....

....

[1 mark]

Undernourishment is down from more than 1 billion in 1992 to around 805 million today, a fall of 21 percent:

161 million children are stunted due to chronic malnutrition

99 million children are underweight

99 million children wasted due to acute malnutrition *WHO*, 2013

3.2 Using **Figure 8**, describe the distribution of undernourishment by world region.

Eastern Asia 161 million

Southern Asia 276 million

Latin America and the Caribbean 37 million

Sub-Saharan Africa 214 million

117 million in other regions

N

Figure 8

....

....

....

....

[2 marks]

3.3 Explain why investing in better nutrition is a way of increasing the chances of a country's development.

....

....

....

....

....

....

....

....

....

....

[6 marks]

Study **Figure 9**, a graph showing the UK's recent production of primary fuels from its own resources.

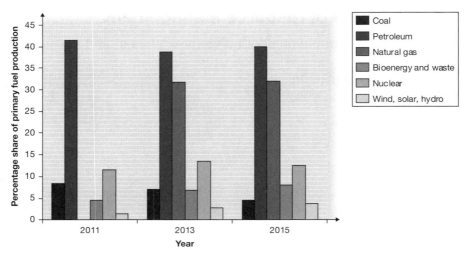

Figure 9

3.4   Using **Figure 9**, complete the graph for natural gas in 2011 using the following data:

2011 Natural gas 33%

[1 mark]

3.5   State the increase in percentage share of primary fuel production of bioenergy and waste between 2011 and 2015.

Shade **one** circle only.

A 3.5%   ◯   C 6.7%   ◯

B 4.4%   ◯   D 7.9%   ◯

[1 mark]

Study **Figure 10**, a graph showing the change in UK electricity generation from renewable sources.

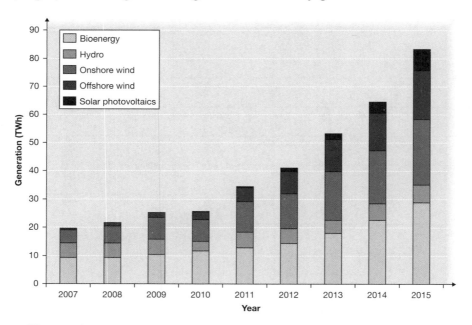

Figure 10

3.6   Using **Figure 10**, suggest why the use of renewable energy sources in the UK is increasing.

....

....

....

....

....

....

[3 marks]

Answer either Question 4 (Food) or Question 5 (Water) or Question 6 (Energy).

Shade the circle below to indicate which optional question you will answer.

Question 4 ◯   Question 5 ◯   Question 6 ◯

Question 4: Food

Study **Figure 11**, a flow line map showing the value of exports of agricultural products from the European Union (EU) in 2014.

Figure 11

4.1   Using **Figure 11**, what was the approximate value of agricultural exports from the EU to China in 2014?

Shade **one** circle only.

A 16.4 billion euros   ◯   **C** 7.2 billion euros   ◯

B 9.1 billion euros   ◯   **D** 5.3 billion euros   ◯

[1 mark]

4.2   Using **Figure 11**, describe the pattern of world agricultural exports from the EU in 2014.

....

....

....

....

[2 marks]

4.3   Give **two** reasons why permaculture is a sustainable way of producing food.

Reason 1: ....

....

Reason 2: ....

....

[2 marks]

4.4   Explain how large-scale agricultural developments can be successful in reducing food insecurity.

....

....

....

....

....

....

....

....

....

....

....

....

[6 marks]

Question 5: Water

Study **Figure 12**, a map showing the amount of water stress in the world.

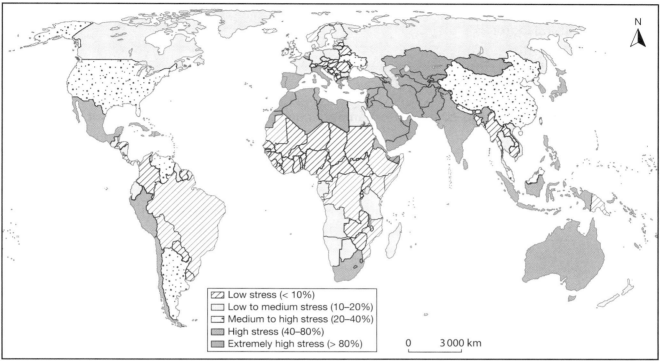

Figure 12

5.1   What was the water stress level in the USA?

Shade **one** circle only.

A Low to medium stress   ◯

B Medium to high stress   ◯

C High stress   ◯

D Extremely high stress   ◯

[1 mark]

5.2   Using **Figure 12**, describe the distribution of water stress in South America.

....

....

....

....

[2 marks]

5.3   Give **two** natural factors that may affect the water supply in a country.

Natural factor 1: ....

....

Natural factor 2: ....

....

[2 marks]

5.4   Explain how large-scale water transfer schemes can help to reduce water insecurity.

....

....

....

....

....

....

....

....

....

....

....

[6 marks]

Question 6: Energy

Study **Figure 13**, a map showing the percentage share of wind and solar energy of total energy production in 2015.

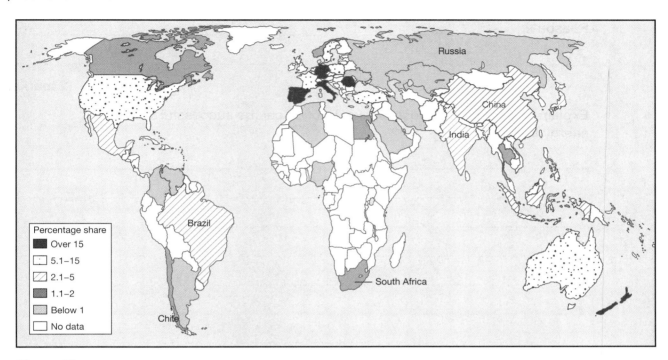

Figure 13

6.1   What was the share of wind and solar energy production in India?

Shade **one** circle only.

A 1.1 to 2%   ◯

B 2.1 to 5%   ◯

C 5.1 to 15%   ◯

D over 15%   ◯

[1 mark]

6.2   Using **Figure 13**, describe the distribution of wind and solar energy production share in South America.

....

....

....

....

[2 marks]

6.3   Give **two** reasons why wind energy is a sustainable way of producing energy.

Reason 1: ....

....

Reason 2: ....

....

[2 marks]

6.4   Explain how large-scale fossil fuel extraction can be successful in reducing energy insecurity.

....

....

....

....

....

....

....

....

....

....

....

[6 marks]

Resources for Paper 3 Geographical applications: Issue evaluation

To be issued to students 12 weeks before the date of the exam.

This booklet contains four resources as follows:

- **Figure 2** – Trends and patterns in air traffic
- **Figure 3** – Reasons for trends and patterns in UK air traffic
- **Figure 5** – Challenges arising from the trends and patterns in UK air traffic
- **Figure 6** – Possible solutions to the South-East airport capacity problem

Figure 2

Trends and patterns in air traffic

According to World Bank data:

- In 1970, the number of world air transport passengers was fewer than 0.5 billion and the UK's share fewer than 20 million.
- As the number of airports and aircraft has increased, so have passenger numbers. By 2015, the number for the world was nearly 3.5 billion and for the UK was more than 130 million.
- A number of NEE countries have built impressive large airports such as Shanghai and Hong Kong in China, Singapore, and Dubai in the United Arab Emirates (UAE).
- Patterns have been affected by the 2008 economic recession and because the latest generation of aircraft are larger and more fuel-efficient.

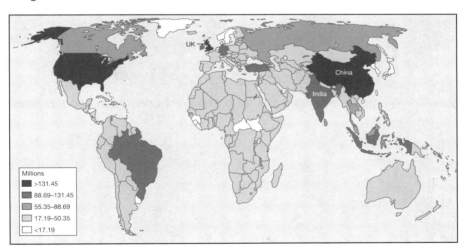

Figure 2.1 Air transport passengers carried by country, 2015

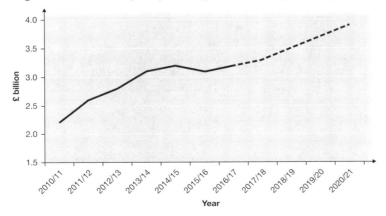

Figure 2.2 UK revenues from Air Passenger Duty (tax) 2010–16 and predictions for the future

Figure 2 continued

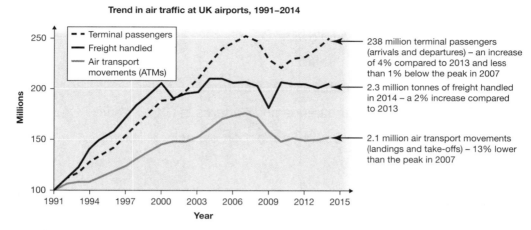

Trend in air traffic at UK airports, 1991–2014

238 million terminal passengers (arrivals and departures) – an increase of 4% compared to 2013 and less than 1% below the peak in 2007

2.3 million tonnes of freight handled in 2014 – a 2% increase compared to 2013

2.1 million air transport movements (landings and take-offs) – 13% lower than the peak in 2007

Figure 2.3 Recent UK transport statistics for aviation

Figure 3

Reasons for trends and patterns in UK air traffic

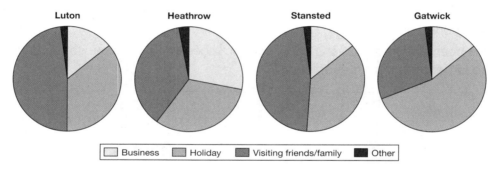

Luton   Heathrow   Stansted   Gatwick

☐ Business   ☐ Holiday   ☐ Visiting friends/family   ■ Other

Figure 3.1 The purpose of air travel for passengers using main London airports in 2015

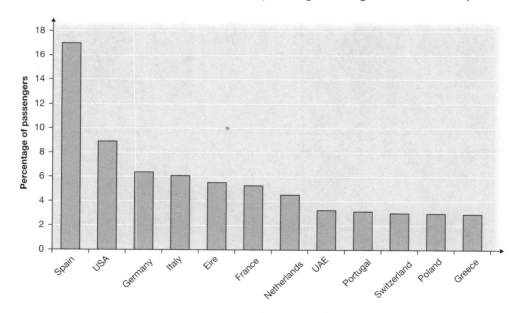

Figure 3.2 UK top 12 arrival and destination countries for air passengers in 2015

Figure 3 continued

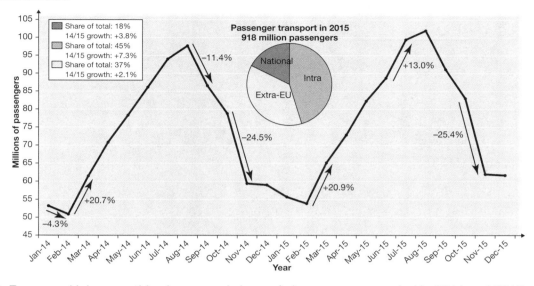

Figure 3.3 European Union monthly change and share of air passengers carried in 2014 and 2015

Figure 5

Challenges arising from the trends and patterns in UK air traffic

Figure 5.1 A 1: 50 000 Ordnance Survey map extract of the Heathrow airport area

Gatwick airport

Heathrow airport

Figure 5.2 Noise patterns (in isobels) for Gatwick and Heathrow airports in 2014

The UK aviation situation

Increasing demand

In 2015, Heathrow was the EU's busiest airport, Gatwick was 8th, Stansted 20th and Luton 30th, with the UK's smaller airports increasing their passenger traffic at a faster rate than both Heathrow and Gatwick. As demand grows, the airports will fill until they reach the limits of either terminal or runway capacities. Over 90 per cent of national passenger capacity is expected to be reached by the early 2040s. Forecasts suggest that all the airports in the South-East will be at capacity by 2030 and the larger airports outside the South-East from 2040. The first airports to fill by 2020 will be in the London area, especially Heathrow and Gatwick.

The UK government commissioned a report to recommend where to expand airport capacity in the South-East of the UK. Three options are available:

1   Expand Heathrow airport to the west of London.

2   Expand Gatwick airport to the south of London.

3   Build a new Thames Hub airport to the east of London.

Figure 6

Possible solutions to the South-East airport capacity problem

Expand Heathrow: benefits for the UK	Only Gatwick can guarantee:
1 It will create over 120000 new jobs, 42% of these outside the London area. 2 As a hub airport, Heathrow is connected to the rest of the UK and has experience with handling lots of goods. 3 It can double its cargo handling capacity, helping 100000 more UK companies trade abroad. 4 It can help meet the government target of doubling exports. 5 New airlines will increase competition and so reduce passenger fares. 6 It will be easier for people to travel abroad or within the UK by air.	1 **Growth across the UK**: we will create the same economic boost as Heathrow, but a bigger boost to regional connectivity across the UK. 2 **A cap on passenger charges** – we will cap air passenger charges at an inflation-linked £15. 3 **No public subsidy**: we will fund the scheme privately and in full. 4 **Legal air quality**: we will not breach legally binding air quality limits with the new runway. 5 **A cap on those most affected by noise**: we will limit the area of the noise contour most affected by noise to $70\,km^2$. 6 **Industry-leading compensation**: we will pay £1000 per annum towards the Council Tax of those most affected by noise.

Figure 6.1 Arguments for Heathrow airport expansion

Figure 6.2 Arguments for expanding Gatwick airport

Thames Hub Airport proposal

Reaction to the Airport Commission final report by Foster and Partners

01 July 2015

The only long-term answer to the question posed to the Airports Commission, 'how do we maintain our global aviation hub status?' is a brand new 4-runway 24-hour airport to the east of the capital, the Thames Hub. It is a bold, flexible, future-proof solution which neither Heathrow nor Gatwick will ever achieve.

The expansion of London towards the east has now been reinforced with the approval of Ebbsfleet Garden City and moves toward a Lower Thames crossing – further validating the relocation of the airport to the Isle of Grain. The argument against the relocation of the bird habitat has been diminished with the success of the Wallasea wetlands project further down the estuary.

The Thames Hub airport will deliver the global connectivity that Londoners demand whilst bringing respite to all tormented by aircraft noise. Connecting the UK to new global destinations throughout the day and night, it can be built for a cost of £26 billion including upgraded road and rail services in just 7 years – the same time it will take for a new runway at Heathrow to be squeezed into a London borough. An integrated rail hub will connect services to Europe, London in just 30 minutes and the rest of the UK.

The case for the Thames Hub Airport

- The Thames Hub Airport will be linked to the rest of the UK and Europe by high-speed rail with a 30-minute ride to central London.
- It will be closer to shipping ports for the efficient distribution of goods.
- An integrated rail hub at the terminal will serve 300000 people a day and cater for the 150-million passengers-per-annum capacity, new 24-hour airport, which will connect the UK to all the emerging economic capitals of the world.
- Air and noise pollution is a quality of life issue and with the capital's population due to expand by 20 per cent over the next two decades, this has become even more pressing. We need to move our hub airport out of our suburbs and unlock the space – equivalent to a new borough – for more homes and jobs within the city.
- The capital is moving east, and the Thames Hub Airport is the obvious answer to the question 'how do we maintain our global aviation hub status?'
- Independent research has shown that in the next decade we need to connect to 55 new destinations across the world to maintain our access to a consistent share of global trade.

Paper 3 Section A: Issue evaluation

Answer all questions in this section.

Study **Figure 1**, a graph showing UK aircraft landings and take-offs between 1970 and 2015. (Note that this graph uses an index based on 1950 = 100).

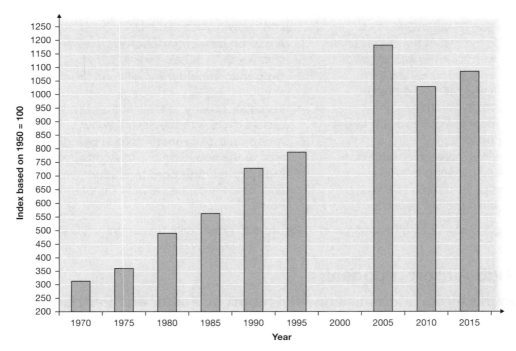

Figure 1

1.1   Using **Figure 1**, complete the graph using the following data for UK aircraft landings and take-offs in the year 2000.

Index for UK aircraft landings and take-offs in 2000 = 1006   **[1 mark]**

Study **Figure 2**, 'Trends and patterns in air traffic', in the resources (pages 299–300).

1.2   With the help of **Figure 2.1**, 'Air transport passengers carried by country, 2015', which **one** of the following statements is correct?

Shade one circle only.

A The UK was the only country with over 131.45 million passengers.   ◯

B India and China were in the same passenger number category.   ◯

C All the countries in South America had between 50.35 and 88.69 million passengers.   ◯

D The continent of Africa had no country with over 50.35 million passengers.   ◯

[1 mark]

1.3   With the help of **Figure 2**, suggest the benefits that the UK may gain from high volumes of air traffic.

....

....

....

....

....

....

....

....

....

....

[6 marks]

Study **Figure 3.3**, 'European Union monthly change and share of air passengers carried in 2014 and 2015', in the resources (page 301).

2.1   In which month of 2014 and 2015 was the number of air passengers in the European Union (EU) at the lowest level?

....

[1 mark]

Study **Figure 3**, 'Reasons for trends and patterns in UK air traffic', in the resources (pages 300–301).

2.2   'Holiday air travel has a major influence on the annual pattern of air traffic in the UK.'

Use **Figure 3** and your own understanding to discuss this statement.

....

....

....

....

....

....

....

....

....

[6 marks]

Study **Figure 4**, a line graph showing carbon dioxide (CO_2) emissions by UK aviation between 1970 and 2020.

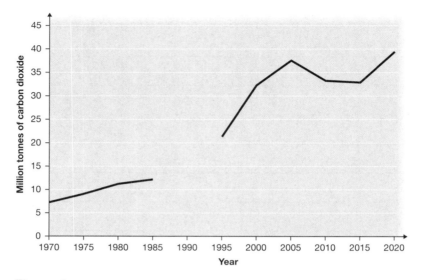

Figure 4

3.1   Using **Figure 4**, complete the graph using the following data for carbon dioxide emissions in 1990.

Million tonnes of carbon dioxide emissions in 1990 = **17.0**

[1 mark]

Study **Figure 5.1**, 'A 1: 50 000 Ordnance Survey map extract of the Heathrow airport area', in the resources (page 301).

3.2   Which **one** of the following six-figure grid references is correct for the bus station in the middle of Heathrow airport?

Shade **one** circle only.

A   050759

B   073759

C   075758

D   085768

[1 mark]

Study **Figure 5.1**, the OS map extract of the Heathrow airport area.

3.3   Suggest **two** reasons why Heathrow is a suitable site for a major UK airport.

Reason 1: ....

....

Reason 2: ....

....

[2 marks]

Study **Figure 5**, 'Challenges arising from the trends and patterns in UK air traffic', in the resources (pages 301–2).

3.4   Using **Figures 5.1** and **5.2**, suggest why people living near Gatwick and Heathrow airports may experience problems due to the trends and patterns in UK air traffic.

....

....

....

....

....

....

....

....

....

....

....

[6 marks]

Study **Figure 6**, 'Possible solutions to the South-East airport capacity problem', in the resources (page 303).

3.5   Three possible locations for expanding the South-East airport capacity have been suggested – Heathrow, Gatwick and the Thames Hub. These are explained in **Figure 6**.

Which one of these three locations do you think will be the most beneficial for the South-East and the UK as a whole?

Use evidence from the resources booklet and your own understanding to explain why you have reached this decision.

[9 marks + 3 SPaG marks]

Use a separate sheet of paper to write your answer to this question. State your chosen location at the start of your answer.

Paper 3 Section B: Fieldwork

Answer all questions in this section.

Study **Figure 7**, a photograph of part of Slough, Berkshire, and **Figure 8**, a photograph of part of the town of Windsor, Berkshire.

Figure 7 Part of Slough, Berkshire

Figure 8 Part of the town of Windsor, Berkshire

4.1   A student was investigating the contrasting deprivation levels in Slough and Windsor. Suggest a possible question for their geographical enquiry.

....

....

[1 mark]

Figure 9 shows data collected from a five-minute traffic count at four different locations in Slough and Windsor.

Slough	
Location 1	92
Location 2	160

Windsor	
Location 3	55
Location 4	15

Figure 9

4.2   Describe one of the possible risks of carrying out a traffic count survey.

....

....

[1 mark]

4.3   State **one** reason why the traffic count data may not be accurate.

....

....

[1 mark]

Figure 10 is a bar chart showing the results from the traffic count.

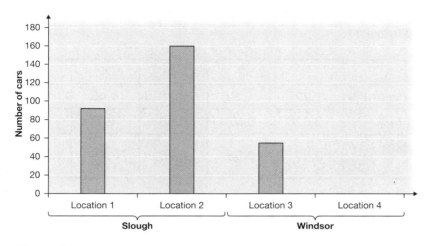

Figure 10

4.4   Complete the bar chart for Location 4 in Windsor on **Figure 10**.

[1 mark]

4.5   Compare the results of the traffic counts for Slough and Windsor.

....

....

....

....

[2 marks]

4.6   Suggest reasons for the results of the traffic counts for Slough and Windsor.

....

....

....

....

[2 marks]

As part of an enquiry collecting primary physical geography data, a student measured the width and depth of the River Horner in Somerset at six different sites, working downstream.

The results are shown in **Figure 11**.

Site	1	2	3	4	5	6
Channel width (m)	1.22	1.19	3.28	4.45	5.08	8.01
Mean depth (m)	0.03	0.07	0.08	0.1	0.22	0.17
Cross-sectional area (m²)	0.04	0.08	0.26	0.45	1.12	

Figure 11

The cross-sectional area of a river is calculated using the following formula:

channel width × mean depth = cross-sectional area

4.7   Using the data in **Figure 11**, calculate the cross-sectional area of the River Horner for Site 6.

Show your working in the space below.

....

....

....

[2 marks]

4.8   Complete the scatter graph below using the data for the cross-sectional area for Site 6.

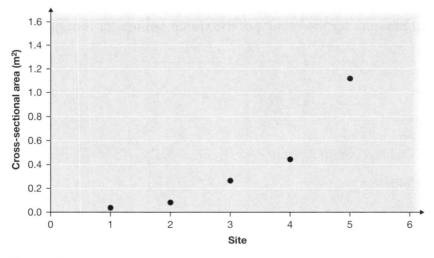

Figure 12

[1 mark]

4.9   Describe **one** of the possible risks of measuring the depth of a river.

....

....

[1 mark]

4.10   Describe the cross-sectional data shown on the scatter graph in **Figure 12**.

....

....

....

....

....

....

[4 marks]

5.1   State the title of your fieldwork enquiry in which **human** geography data was collected.

Title of fieldwork enquiry:

....

Describe **one** of the primary data collection methods used for your fieldwork enquiry.

....

....

....

[2 marks]

5.2   Explain how you presented the primary data that you collected for your human geography fieldwork.

....

....

....

....

[3 marks]

5.3   State the title of your fieldwork enquiry in which physical geography data was collected.

Title of fieldwork enquiry:

....

On a separate sheet of paper assess how successful your primary data collection was in helping you complete your fieldwork enquiry.

[6 marks]

5.4   For **one** of your geography enquiries, explain the results and conclusions of your investigation.

[9 marks + 3 SPaG marks]

Use a separate sheet of paper to write your answer to this question.

Answers

The challenge of natural hazards: Tectonic hazards

Question	Marking guidance	Total marks
1	Two points made: • earthquakes are found along plate boundaries • for example, close to the western edge of the South American plate.	2
2	One mark for a reason and the second mark for an extension. • Earthquakes happen where tectonic plates meet. • For example, a destructive plate boundary, where an oceanic plate is subducted underneath a continental plate.	2
3	**Indicative content** • The focus of the question must be on volcanoes at destructive plate margins. • A destructive plate margin is where two plates are moving towards each other. • The denser oceanic plate is subducted beneath the lighter continental plate. • As the oceanic plate moves downwards it melts due to friction and heat from the mantle. • Less dense magma is created which breaks through to the surface to form steep-sided composite volcanoes. • The volcanic eruptions are often very violent and explosive. **Level 2 (Clear) 3–4 marks:** • AO2 Shows clear geographical understanding of why volcanoes occur at destructive plate boundaries. • AO3 Demonstrates application of knowledge and understanding to make full interpretation of the processes which cause volcanoes to occur at destructive plate boundaries. **Level 1 (Basic) 1–2 marks:** • AO2 Shows limited geographical understanding of why volcanoes occur at destructive plate boundaries. • AO3 Demonstrates application of knowledge and understanding to make limited interpretations of the processes which cause volcanoes to occur at destructive plate boundaries. • 0 marks: No relevant content.	4
3	**Indicative content** • Buildings and bridges can be constructed to resist the ground shaking associated with an earthquake. • Walls can be reinforced with steel and concrete to reduce movement. • Shock absorbers can absorb ground shaking. • There are open areas where people go for easy evacuation. • Automatic shutters can come down over windows to prevent broken glass falling. • Rolling weights on a roof can counteract seismic/shock waves. **Level 2 (Clear) 3–4 marks:** • AO1 Demonstrates detailed knowledge of how people can be protected from earthquakes. • AO2 Shows clear geographical understanding of why people need to be protected from earthquakes. **Level 1 (Basic) 1–2 marks:** • AO1 Demonstrates knowledge of how people can be protected from earthquakes. • AO2 Shows some geographical understanding of why people need to be protected from earthquakes. • 0 marks: No relevant content.	4
4	**Indicative content** • Primary effects: number of people killed/injured; buildings destroyed, e.g. schools and hospitals; homelessness; electricity, water supplies, sanitation and communications affected; need for food and shelter. • Secondary effects: landslides; flooding; spread of disease, e.g. cholera; tsunami; fires. • Named example: should be both place and date specific, e.g. the L'Aquila earthquake in central Italy on 6 April 2009. **Level 3 (Detailed) 5–6 marks:** • AO3 Demonstrates thorough application of knowledge and understanding to give detailed description of earthquakes. Includes detailed place-specific information. • AO3 Shows full understanding of the interrelationships between primary and secondary effects, using evidence to support response. **Level 2 (Clear) 3–4 marks:** • AO1 Demonstrates specific and accurate knowledge of the effects of an earthquake. Includes place-specific information. • AO2 Shows clear geographical understanding of primary and secondary effects. **Level 1 (Basic) 1–2 marks:** • AO1 Demonstrates some knowledge of the effects of earthquakes. May include some place-specific information. • AO2 Shows limited geographical understanding of difference between primary and secondary effects. • 0 marks: No relevant content.	6

The challenge of natural hazards: Weather hazards

Question	Marking guidance	Total marks
1	One mark for each correct answer: **B** Tropical storms form above warm oceans (27°C or above). **D** The conditions in the eye of the storm are calm. No credit if three or more statements are shaded.	2
2	One mark for a reason and the second mark for an extension: • a large number of people die by drowning in a storm surge • people are made homeless due to strong winds damaging homes.	2
3	**Indicative content** • Windows, doors and roofs reinforced to strengthen buildings to withstand strong winds. • Houses constructed on stilts so that a storm surge will pass beneath. • Storm shelters built. • Educating people to be prepared for tropical storms. **Level 2 (Clear) 3–4 marks:** • AO1 Demonstrates detailed knowledge of how people can prepare for tropical storms. • AO2 Shows clear geographical understanding of why people need to prepare for tropical storms. **Level 1 (Basic) 1–2 marks:** • AO1 Demonstrates knowledge of how people can prepare for tropical storms. • AO2 Shows some geographical understanding of why people need to prepare for tropical storms. • 0 marks: No relevant content.	4

| 4 | **Indicative content**
• Primary effects: number of people killed; people displaced/homeless; homes destroyed; buildings destroyed, e.g. schools and hospitals; communications/transport links interrupted and/or destroyed; fishing boats destroyed; crops destroyed; flooding.
• Secondary effects: loss of jobs and income; flooding causes landslides, cutting off roads; power supplies cut off; shortages of water, food and shelter, which can lead to outbreaks of disease; education disrupted after schools are destroyed; looting of shops.
• Named example: should be both place and date specific, e.g. Typhoon Haiyan, Philippines, 2013.

Level 3 (Detailed) 7–9 marks:
• AO1 Demonstrates detailed knowledge of the primary and secondary effects of a tropical storm.
• AO2 Shows thorough geographical understanding of the interrelationships between places, environments and processes in the context of tropical storms.
• AO3 Demonstrates application of knowledge and understanding in an effective and reasoned way in assessing the significance of effects.

Level 2 (Clear) 4–6 marks:
• AO1 Demonstrates clear knowledge of the primary and secondary effects of a tropical storm.
• AO2 Shows some geographical understanding of the interrelationships between places, environments and processes in the context of tropical storms.
• AO3 Demonstrates reasonable application of knowledge and understanding in assessing the significance of effects.

Level 1 (Basic) 1–3 marks:
• AO1 Demonstrates limited knowledge of the primary and secondary effects of a tropical storm.
• AO2 Shows slight understanding of the interrelationships between places, environments and processes in the context of tropical storms.
• AO3 Demonstrates limited application of knowledge and understanding in assessing the significance of effects.
• 0 marks: No relevant content. | 9+3 SPaG |

Spelling, punctuation and grammar		
High performance	3 marks	• Spelling and punctuation are consistently accurate. • Answers use rules of grammar with effective overall control of meaning. • Answers use a wide range of specialist terms as appropriate.
Intermediate performance	2 marks	• Spelling and punctuation are generally accurate. • Answers use rules of grammar with general overall control of meaning. • Answers use a good range of specialist terms as appropriate.
Threshold performance	1 mark	• Spelling and punctuation are reasonably accurate. • Answers use rules of grammar with some control of meaning and any errors do not significantly hinder meaning overall. • Answers use a limited range of specialist terms as appropriate.
Below threshold performance	0 marks	• Nothing is written. • The answer does not relate to the question. • The answer SPaG does not reach the threshold performance level, for example, errors in spelling, punctuation and grammar severely hinder meaning.

| 5 | **Indicative content**
Social examples, such as:
• houses flooded
• residents evacuated and put into temporary shelters
• farms evacuated
• daily routine affected – going to work, school, etc.
• power supplies cut off.
Economic examples, such as:
• cost of flood damage
• agricultural land affected
• livestock evacuated
• local roads cut off affecting people's journeys to work
• railway lines affected.
Environmental examples, such as:
• floodwaters contaminated with sewage and other pollutants
• debris deposited which had to be cleared
• stagnant water – had to be reoxygenated before putting back into a river.
• Named example: should be both place and date specific, e.g. the Somerset floods of January 2014.

Level 3 (Detailed) 5–6 marks:
• AO1 Demonstrates detailed knowledge of the impacts of a flood on the people and environment.
• AO2 Shows thorough geographical understanding of the interrelationship between places, environments and processes in the context of flooding.
• AO3 Demonstrates application of knowledge and understanding in assessing the significance of the impacts.

Level 2 (Clear) 3–4 marks:
• AO1 Demonstrates clear knowledge of the impacts of a flood on the people and environment.
• AO2 Shows some geographical understanding of the interrelationship between places, environments and processes in the context of flooding.
• AO3 Demonstrates reasonable application of knowledge and understanding in assessing the significance of the impacts.

Level 1 (Basic) 1–2 marks:
• AO1 Demonstrates limited knowledge of the impacts of a flood on the people and environment.
• AO2 Shows slight geographical understanding of the interrelationship between places, environments and processes in the context of flooding.
• AO3 Demonstrates limited application of knowledge and understanding in assessing the significance of the impacts.
• 0 marks: No relevant content. | 6 |

| 6 | **Indicative content**
• Answers should include a judgement to the extent they agree with the statement, so answer should provide evidence for and against the statement. Answers must relate to the UK.
• More energy in the atmosphere could lead to more intense storms.
• The atmospheric circulation could be affected, bringing more floods to normally dry regions and heat waves to normally cooler areas.
• No single weather event can be blamed on climate change.
• Evidence to support more extreme weather – 2013/14 and 2015/16 heavy rainfall causing devastating floods; 2009 and 2010 heavy snowfall.

Level 3 (Detailed) 7–9 marks:
• AO1 Demonstrates comprehensive knowledge of how the weather in the UK may change.
• AO2 Shows thorough geographical understanding of the interrelationships between places, environments and processes in the context of the UK's weather.
• AO3 Demonstrates thorough application of knowledge and understanding in evaluating the extent to which the UK's weather is becoming more extreme.

Level 2 (Clear) 4–6 marks:
• AO1 Demonstrates clear knowledge of how the weather in the UK may change.
• AO2 Shows some geographical understanding of the interrelationships between places, environments and processes in the context of the UK's weather.
• AO3 Demonstrates reasonable application of knowledge and understanding in evaluating the extent to which the UK's weather is becoming more extreme.

Level 1 (Basic) 1–3 marks:
• AO1 Demonstrates limited knowledge of how the weather in the UK may change.
• AO2 Shows slight understanding of the interrelationships between places, environments and processes in the context of the UK's weather.
• AO3 Demonstrates limited application of knowledge and understanding in evaluating the extent to which the UK's weather is becoming more extreme.
• 0 marks: No relevant content.
For SPaG marks, please see table above. | 9+3 SPaG |

Answers

The challenge of natural hazards: Climate change

Question	Marking guidance	Total marks
1	One mark for a reason and the second mark for an extension: volcanic ash can block out sunlight. This leads to a reduction in temperatures in the short term.	2
2	One mark each for two of the following: alternative energy sources, carbon capture, planting trees, international agreements.	2
3	**Indicative content** • The enhanced greenhouse effect is the impact caused by the heat retained in the atmosphere by the greenhouse gases added to the atmosphere through human activity. • More carbon dioxide has been added to the atmosphere through burning fossil fuels, increased transportation and deforestation. • Methane has been added to the atmosphere through more intense farming, such as rice farming and cattle ranching. **Level 2 (Clear) 3–4 marks:** • AO1 Demonstrates detailed knowledge of the enhanced greenhouse effect. • AO2 Shows clear geographical understanding of what causes the enhanced greenhouse effect. **Level 1 (Basic) 1–2 marks:** • AO1 Demonstrates knowledge of the enhanced greenhouse effect. • AO2 Shows some geographical understanding of what causes the enhanced greenhouse effect. • 0 marks: No relevant content.	4
4	**Indicative content** • Natural causes of climate change are orbital changes such as eccentricity, which causes changes in the elliptical path of the Sun, affecting the temperature of the Earth. • Increase in solar energy linked to sunspots: the more sunspots there are the more solar radiation is given out. • Volcanic ash can block out sunlight and so temperatures are reduced. • Sulfur dioxide from a volcanic eruption mixes with water to cause sulfuric acid. This reflects the Sun's radiation, reducing temperatures. **Level 2 (Clear) 3–4 marks:** • AO1 Demonstrates detailed knowledge of natural causes of climate change. • AO2 Shows clear geographical understanding of what causes natural climate change. **Level 1 (Basic) 1–2 marks:** • AO1 Demonstrates knowledge of natural causes of climate change. • AO2 Shows some geographical understanding of what causes natural climate change. • 0 marks: No relevant content.	4
5	**Indicative content** • Burning fossil fuels is the largest contributor to climate change. • Fossil fuels are used in transport, industry and power stations to generate electricity. • Intensive agriculture has contributed to an increase in methane due to cattle and rice farming. • Forests act as carbon stores. Deforestation has reduced the number of trees. This means more carbon dioxide is released into the atmosphere. **Level 3 (Detailed) 5–6 marks:** • AO1 Demonstrates detailed knowledge of how humans have contributed to climate change. • AO2 Shows thorough geographical understanding of the interrelationship between places, environments and processes in the context of climate change. **Level 2 (Clear) 3–4 marks:** • AO1 Demonstrates clear knowledge of how humans have contributed to climate change. • AO2 Shows some geographical understanding of the interrelationship between places, environments and processes in the context of climate change. **Level 1 (Basic) 1–2 marks:** • AO1 Demonstrates limited knowledge of how humans have contributed to climate change. • AO2 Shows slight geographical understanding of the interrelationship between places, environments and processes in the context of climate change. • 0 marks: No relevant content.	6

The living world: Ecosystems

Question	Marking guidance	Total marks
1	One mark for the correct answer: **C** Soils get nutrients from the decomposition of litter. No credit if two or more statements are shaded.	1
2	Credit statements about what water does in the nutrient cycle using information in the diagram, such as: • washing of nutrients deep into a soil (leaching) (1) • rain runs over the ground and carries nutrients away (1) • rain supplies some dissolved nutrients (1) • moisture helps the decomposition process (1) • water dissolves nutrients so that plants can absorb them through their roots (1). No credit for simply listing parts of the water or nutrient cycles.	2
3	One mark for the correct answer: **C** Large temperature range during a year (−250°C to +100°C) and low precipitation (approximately 500 mm a year). No credit if two or more statements are shaded.	1
4	**Indicative content** • Responses should include description and explanation. • There should be evidence that the candidate knows the small-scale ecosystem, e.g. types of flora and fauna, soil and climate. • Links should be clear and specific, showing understanding of food webs, nutrient cycling and other processes. • Max level 1 for either description or explanation. **Level 3 (Detailed) 5–6 marks:** • AO1 Applies detailed knowledge of small-scale UK ecosystem to interpret the question. • AO3 Clearly relates characteristics of the ecosystem to links. **Level 2 (Clear) 3–4 marks:** • AO1 Demonstrates accurate knowledge of the features of vegetation, climate and other natural factors to support the explanation. • AO3 Relates some characteristics of the ecosystem to links. • AO4 Makes effective use of diagrams. **Level 1 (Basic) 1–2 marks:** • AO1 Demonstrates limited knowledge of vegetation, climate and other natural factors in UK ecosystem. • AO4 Makes no use of visual material. • 0 marks: No relevant content.	6

5	**Indicative content** • Responses should include description *and* explanation. • There should be evidence that the candidate knows the small-scale ecosystem, e.g. the links between different parts and between factors influencing change. • Climate/weather extremes causing change. • Human management of the ecosystem. • Conflict between nature and specified human activities or presence. • Max level 1 for either description or explanation. **Level 3 (Detailed) 5–6 marks:** • AO1 Applies detailed knowledge of small-scale UK ecosystem to interpret the question. • AO3 Clearly relates characteristic(s) of the ecosystem to factors affecting the natural balance. **Level 2 (Clear) 3–4 marks:** • AO1 Demonstrates accurate knowledge of the features of the UK ecosystem to support the explanation. • AO4 Makes clear and effective use of diagrams. **Level 1 (Basic) 1–2 marks:** • AO1 Demonstrates limited knowledge of the UK ecosystem and the balance within it. • AO4 Makes limited use visual material • 0 marks: No relevant content.	6
6	**Indicative content for tropical rainforests, hot deserts and/or cold environments (as appropriate)** • The question requires consideration of the extent to which the interactions between the living and non-living parts of two global ecosystems create a balance in the ecosystem. • Answers may describe the balance that exists in two or more global ecosystems. • Answers may focus on specific examples within generalisations such as climate, which determines plant growth (producers) at the base of trophic levels through heat, rain and sunlight; soils providing nutrients for plant growth and humus being added by plants (nutrient cycle); the supply of water for plants and animals. • Answers may consider factors that upset the balance such as climate change and human actions. • Answers must consider natural processes. **Level 3 (Detailed) 7–9 marks:** • AO1 Demonstrates comprehensive and accurate knowledge of locations, places and processes in relation to tropical rainforests, hot deserts and/or cold environments. • AO2 Shows thorough geographical understanding of the interrelationships between places, environments and processes in the context of tropical rainforests, hot deserts and/or cold environments. • AO3 Demonstrates thorough application of knowledge and understanding in evaluating the extent to which interactions between biotic and abiotic parts create balance in tropical rainforests, hot deserts and/or cold environments. **Level 2 (Clear) 4–6 marks:** • AO1 Demonstrates clear knowledge of locations, places and processes in relation to tropical rainforests, hot deserts and/or cold environments. • AO2 Shows some geographical understanding of the interrelationships between places, environments and processes in the context of tropical rainforests, hot deserts and/or cold environments. • AO3 Demonstrates reasonable application of knowledge and understanding in evaluating the extent to which interactions between biotic and abiotic parts create balance in tropical rainforests, hot deserts and/or cold environments. **Level 1 (Basic) 1–3 marks:** • AO1 Demonstrates limited knowledge of locations, places and processes in relation to tropical rainforests, hot deserts and/or cold environments. • AO2 Shows slight geographical understanding of the interrelationships between places, environments and processes in the context of tropical rainforests, hot deserts and/or cold environments. • AO3 Demonstrates limited application of knowledge and understanding in evaluating the extent to which interactions between biotic and abiotic parts create balance in tropical rainforests, hot deserts and/or cold environments. • 0 marks: No relevant content. **For SPaG marks, please see table in Weather hazards section.**	9+3 SPaG

The living world: Tropical rainforests		
Question	**Marking guidance**	**Total marks**
1	One mark for the correct answer: **D** Nutrient cycling is very fast. No credit if two or more statements are shaded.	1
2	**Indicative content** • An area of tropical rainforest must be identified. • Responses should include description *and* explanation. • There should be evidence that the candidate knows the global ecosystem and what has been happening to it. • Causes such as: LIC/NEEs trying to make money to help development; businesses trying to make money; poor people trying to improve their lives; country trying to use resources. • Max level 1 for either description or explanation. **Level 3 (Detailed) 5–6 marks:** • AO1 Applies detailed knowledge of a tropical rainforest to interpret the question. • AO3 Clearly relates a description of deforestation to reasons for the deforestation. **Level 2 (Clear) 3–4 marks:** • AO1 Demonstrates accurate knowledge of the features of deforestation to support the explanation. • AO4 Makes clear and effective use of photograph. **Level 1 (Basic) 1–2 marks:** • AO1 Demonstrates limited knowledge of deforestation in a tropical rainforest. • AO4 Makes limited use visual material. • 0 marks: No relevant content.	6
3	One mark for the correct answer: French Guiana. No credit for Guyana or any other country.	1
4	One mark for the correct answer: Four/4 (countries).	1

5	Credit one value only. Value must be to people, not to the natural environment. One mark for stating a value, such as: • source of valuable resources (1) • place to live and make a living (1) • provision of oxygen (1) • store carbon (1) • biodiversity (1). Second mark for a developed explanation. • Resources in the forest such as foods, timber, medicines or under the forest, such as coal, oil, metals, can be used to provide jobs and make money for people and countries (1). • As populations grow, people need living space and ways of farming to get their own food or products to sell (1). • People need oxygen to breathe, which is produced by plants and the greatest number of trees is in the tropical rainforest (1). • Climate change is and will continue to greatly affect people; the tropical rainforest stores carbon and stops some CO_2 getting into the atmosphere to create more warming (1). • The huge biodiversity provides chemicals and substances which people may be able to use to create medicines and other beneficial products (1).	2
6	Credit one impact only. The answer must focus on a negative impact of deforestation of a tropical rainforest and should be referenced to one area (e.g. Malaysia or South-East Asian rainforest). One mark for stating a negative impact, such as: • reduction of biodiversity (1) • erosion of soils (1) • increased rate of climate change (1) • local area becomes drier (1) or may lead to increased flooding (1) • water and air pollution increase (1) • reduction of eco-tourism (1) Second mark for a developed explanation. • If biodiversity is reduced, the wildlife may be made extinct, creating a loss of a resource to nature (food web unbalanced) and people (source of medicine) (1). • Erosion and degradation of soils means that the forest or crops will not grow and natural resources will be lost (1). • Climate change threatens natural ecosystems by making them hotter or drier, which then reduces resources to support people (1). • Drier climate damages trees and causes dieback and further forest loss (1). • Pollution damages the conditions in which trees and plants live and also the health of people (1). • A damaged forest does not appeal to people for their holidays and so tourism income is lost (1).	2
7	Credit one way only. The answer must focus on education of people and sustainability of the tropical rainforest. One mark for stating an aspect of education, such as: • value of the tropical rainforest (1) • ways of looking after tropical rainforest (1) • rates and types of damage (1) • ways of reducing damage to tropical rainforest (1) • alternative ways of making money (1). Second mark for a developed explanation. • If people appreciate the need for biodiversity and the benefits from tropical rainforest products, they are less likely to cut them down (1). • Understanding of the role of national parks, biosphere reserves and nature reserves so that their designation is recognised by people (1). • Understanding of the rapid loss and the consequences for local people, countries and the world would make people appreciate the need to look after tropical rainforests (1). • Sustainable logging practices, agroforestry rather than clearance and shifting cultivation help maintain forest structure and biodiversity (1). • Instead of cutting down tropical rainforests for money, things like eco-tourism can provide income based on an undamaged tropical rainforest (1).	2
8	**Indicative content for tropical rainforests (as appropriate)** • The question requires consideration of the extent to which the ecosystem provides opportunities and challenges for development. • Answers must focus on both opportunities and challenges. Opportunities such as: providing a range of resources, which can be used to provide jobs and money, including exports; providing living space for poorer people to help improve their quality of life; providing opportunities for agroforestry to provide food and export crops. Challenges such as: soils are infertile; resources difficult to access; some resources, such as medical compounds, are undiscovered; a very complex ecosystem, which makes it difficult to assess impacts of human activities; reducing loss of biodiversity; preventing changes to climate balance. • Must indicate, through reference to specific examples, the balance between opportunities and challenges. **Level 3 (Detailed) 7–9 marks:** • AO1 Demonstrates comprehensive and accurate knowledge of locations, places and processes in relation to a tropical rainforest. • AO2 Shows thorough geographical understanding of the interrelationships between places, environments and processes in the context of a tropical rainforest. • AO3 Demonstrates thorough application of knowledge and understanding in evaluating the extent to which the ecosystem provides opportunities and challenges for development. **Level 2 (Clear) 4–6 marks:** • AO1 Demonstrates clear knowledge of locations, places and processes in relation to a tropical rainforest. • AO2 Shows some geographical understanding of the interrelationships between places, environments and processes in the context of a tropical rainforest. • AO3 Demonstrates reasonable application of knowledge and understanding in evaluating the extent to which the ecosystem provides opportunities and challenges for development. **Level 1 (Basic) 1–3 marks:** • AO1 Demonstrates limited knowledge of locations, places and processes in relation to a tropical rainforest. • AO2 Shows slight geographical understanding of the interrelationships between places, environments and processes in the context of a tropical rainforest. • AO3 Demonstrates limited application of knowledge and understanding in evaluating the extent to which the ecosystem provides opportunities and challenges for development. • 0 marks: No relevant content. **For SPaG marks, please see table in Weather hazards section.**	9+3 SPaG

The living world: Hot deserts		
Question	**Marking guidance**	**Total marks**
1	One mark for the correct answer: **C** 29°C No credit if two or more statements are shaded.	1
2	One mark for the correct answer: 7 mm. (Units must be given.)	1
3	Credit statements about causes of desertification, such as: • climate has changed to be drier in some semi-arid areas (1) • overpopulation has led to cutting down too many trees for fuel and using up too much water (1) • overgrazing by livestock has removed grasses that held soil together (1) • higher evaporation rates has removed moisture from soils allowing it to be blown away (1) No credit for simply defining desertification or describing what the area is like.	2

4	**Indicative content**	4
	• Using fences (e.g. brushwood) as windbreaks (1) to slow the movement of sand into semi-arid areas (1).	
	• Terracing fields and slopes (1) so that any rain has time to soak into soils rather than running away (1).	
	• Creating furrows across fields so that any rainwater is trapped and has time to soak into the soils (1).	
	• Plant drought-resistant trees (1) so that they hold soil together, shelter it from wind and shade the land so that evaporation is reduced (1).	
	Level 2 (Clear) 3–4 marks:	
	• AO3 Demonstrates accurate interpretation of desertification through the application of relevant understanding and knowledge.	
	Level 1 (Basic) 1–2 marks:	
	• AO3 Demonstrates some interpretation of desertification through the limited application of relevant understanding and knowledge.	
	• 0 marks: No relevant content.	
5	**Indicative content**	4
	• The lack of water means that even plants that have adapted find it difficult to survive (1) as shown by the dead trees in Figure 6 (1).	
	• Any water or lakes that form may evaporate quickly (1) as shown by the cracked ground in the photo (1).	
	• It is often dry for long periods of time with clear skies (as shown in Figure 6) (1) so plants and animals have difficulty finding shade (1).	
	Level 2 (Clear) 3–4 marks:	
	• AO3 Demonstrates accurate explanation of hot desert ecosystem vulnerability through the application of relevant knowledge and understanding.	
	• AO4 Makes clear and effective use of the image to support the explanation of vulnerability.	
	Level 1 (Basic) 1–2 marks:	
	• AO3 Demonstrates some explanation of hot desert ecosystem vulnerability through the application of limited relevant knowledge and understanding.	
	• AO4 Makes limited use of the image to support the explanation of vulnerability.	
6	**Indicative content**	6
	Responses must include description **and** explanation.	
	There should be evidence that the candidate knows a hot desert ecosystem.	
	Links should be clear and specific, showing understanding of importance, such as:	
	• delicate balance in a fragile environment so links could be easily broken	
	• many animals are nocturnal because that is when it is cooler	
	• plants are adapted to store water when available or have long roots to get to groundwater because there is little rain	
	• plants and insects have adapted their growth cycles to be fast when rain does occur	
	• soils lack nutrients so plants only grow slowly	
	• oases are places of greater biodiversity due to the water supply.	
	Max level 1 for either description or explanation.	
	Level 3 (Detailed) 5–6 marks:	
	• AO1 Applies detailed knowledge of a hot desert ecosystem to interpret the question.	
	• AO3 Clearly relates characteristic(s) of the ecosystem to links and their importance.	
	Level 2 (Clear) 3–4 marks:	
	• AO1 Demonstrates accurate knowledge of how the links between animals, vegetation, climate and other natural factors are important.	
	• AO4 Makes effective use of diagrams.	
	Level 1 (Basic) 1–2 marks:	
	• AO1 Demonstrates limited knowledge of how animals, vegetation, climate and other natural factors are important.	
	• AO4 Makes no use of diagrams.	
	• 0 marks: No relevant content.	
7	**Indicative content**	9+3 SPaG
	The hot desert environment should be stated. In the accompanying AQA GCSE Geography Revision Guide, this is the Thar Desert.	
	The question requires consideration of the extent to which opportunities are greater than the challenges for development; a candidate may decide that they are not greater.	
	Answers must focus on both opportunities and challenges and compare them. Example opportunities are:	
	• extracting valuable minerals (such as gypsum, marble, lignite and oil), which help energy security and give industries materials they can use	
	• clear skies for developing solar energy and strong winds for wind energy – renewable resources providing electricity for people	
	• desert festivals attracting tourist income	
	• large quantities of different crops can be grown once irrigation water is provided.	
	Example challenges are:	
	• extreme heat during the day and in summer making it difficult to work	
	• high evaporation rates contaminating fields with salts (salinisation) and causing large water losses from irrigation canals	
	• population growth placing greater and greater demands on the water supply situation	
	• transport infrastructure is difficult to maintain because of heat and moving sand.	
	Must indicate, through reference to specific examples, whether the opportunities are greater than the challenges or not.	
	Level 3 (Detailed) 7–9 marks:	
	• AO1 Demonstrates comprehensive and accurate knowledge of locations, places and processes in relation to a hot desert environment.	
	• AO2 Shows thorough geographical understanding of the interrelationships between places, environments and processes in the context of a hot desert environment.	
	• AO3 Demonstrates thorough application of knowledge and understanding in evaluating the extent to which the opportunities are greater than the challenges for development.	
	Level 2 (Clear) 4–6 marks:	
	• AO1 Demonstrates clear knowledge of locations, places and processes in relation to a hot desert environment.	
	• AO2 Shows some geographical understanding of the interrelationships between places, environments and processes in the context of a hot desert environment.	
	• AO3 Demonstrates reasonable application of knowledge and understanding in evaluating the extent to which opportunities are greater than the challenges for development.	
	Level 1 (Basic) 1–3 marks:	
	• AO1 Demonstrates limited knowledge of locations, places and processes in relation to a hot desert environment.	
	• AO2 Shows slight geographical understanding of the interrelationships between places, environments and processes in the context of a hot desert environment.	
	• AO3 Demonstrates limited application of knowledge and understanding in evaluating the extent to which opportunities are greater than the challenges for development.	
	• 0 marks: No relevant content.	
	For SPaG marks, please see table in Weather hazards section.	

The living world: Cold environments		
Question	**Marking guidance**	**Total marks**
1	One mark for the correct answer: **D** –9.9°C No! If the temperature is taken as the middle of December column, then the average temp is –11.5°C No credit if two or more statements are shaded AO4 = 1 mark	1
2	One mark for the correct answer: 20 mm. (Units must be given.)	1

3	Credit statements about adaptations to the cold or dry conditions, such as: • rapid plant growth in short summer (1) • plants grow low to ground to avoid wind-chill effect (1) • white camouflage used by herbivores and carnivores to avoid being spotted against the snow (1) • caribou have wide hooves to avoid sinking into waterlogged ground (1) • many animals have layers of fur or fat to keep warm (1). No credit for simple listing, candidates must outline.	2
4	**Indicative content** • The lack of liquid water for part of a year means that even plants that have adapted find it difficult to survive (1) as shown by the bare patches of ground in Figure 8 (1). • Snow and ice can cover the ground for a long time making it difficult for plants to grow through (1) as shown by the large patch of snow in the photo (1). • The cold conditions are made worse by wind chill so plants have to grow close to the ground to survive (1) as shown by the low plants in the foreground of Figure 8 and the absence of trees (1). **Level 2 (Clear) 3–4 marks:** • AO3 Demonstrates accurate explanation of cold environment ecosystem vulnerability through the application of relevant knowledge and understanding. • AO4 Makes clear and effective use of the image to support the explanation of vulnerability. **Level 1 (Basic) 1–2 marks:** • AO3 Demonstrates some explanation of cold environment ecosystem vulnerability through the application of limited relevant knowledge and understanding. • AO4 Makes limited use of the image to support the explanation of vulnerability.	4
5	One mark for each suggestion (up to a maximum of two) that shows understanding of the benefits of cold wilderness areas for people, such as: • genetic diversity for scientific study that may give future medicines • training areas for isolation in space exploration • past and present climate change information from ice cores in Antarctica, etc. • provides comparison with areas that have been damaged by human activities. No credit for vague statements such as 'see animals' or 'go on holiday to'.	2
6	**Indicative content** Responses must include description and explanation. There should be evidence that the candidate knows cold environment locations. Methods should be clear and specific, showing understanding, such as: • design oil pipelines to withstand the climate conditions and the permafrost (e.g. the Alaskan pipeline) • establish nature reserves or national parks to protect areas (e.g. Arctic National Wildlife Reserve in Alaska, USA) • obey international treaties established to protect areas and wildlife (e.g. Antarctic Treaty and ban on whaling) • develop new international agreements to protect areas (e.g. in the Arctic). Max level 1 for either description or explanation. **Level 3 (Detailed) 5–6 marks:** • AO1 Applies detailed knowledge of a cold environment to interpret the question. • AO3 Clearly relates problems of economic development to how they can be reduced. **Level 2 (Clear) 3–4 marks:** • AO1 Demonstrates accurate knowledge of a cold environment and links it to problems of economic development. • AO3 Makes effective use of understanding of how to reduce problems. **Level 1 (Basic) 1–2 marks:** • AO1 Demonstrates limited knowledge of a cold environment that has experienced economic development. • AO3 Makes no reference to solving or reducing problems. • 0 marks: No relevant content.	6
7	**Indicative content** The question requires consideration of the extent to which climate determines the characteristics of a cold environment ecosystem. The answer must indicate, through reference to specific examples, how the natural environment is affected by climate and suggest which other factors may be important. Answers may focus on: • very low temperatures in Antarctica that create an ice sheet so wildlife is only found at the fringes of the continent • animals and plants have to adapt to the cold through having layers of fat or fur to keep warm • the cold temperatures freeze the ground (permafrost) so that it is difficult for plants to grow or animals to get food • widespread snow and ice has encouraged animals to camouflage themselves • plants are small to escape the cold winds • human activities affected by the climate. **Level 3 (Detailed) 7–9 marks:** • AO1 Demonstrates comprehensive and accurate knowledge of locations, places and processes in relation to a cold environment. • AO2 Shows thorough geographical understanding of the interrelationships between climate and biotic and abiotic parts of a cold environment ecosystem. • AO3 Demonstrates thorough application of knowledge and understanding in evaluating the extent to which the ecosystem characteristics result from the climate. **Level 2 (Clear) 4–6 marks:** • AO1 Demonstrates clear knowledge of locations, places and processes in relation to a cold environment. • AO2 Shows some geographical understanding of the interrelationships between climate and biotic and abiotic parts of a cold environment ecosystem. • AO3 Demonstrates reasonable application of knowledge and understanding in evaluating the extent to which the ecosystem characteristics result from the climate. **Level 1 (Basic) 1–3 marks:** • AO1 Demonstrates limited knowledge of locations, places and processes in relation to a cold environment. • AO2 Shows slight geographical understanding of the interrelationships between climate and biotic and abiotic parts of a cold environment ecosystem. • AO3 Demonstrates limited application of knowledge and understanding in evaluating the extent to which the ecosystem characteristics result from climate. • 0 marks: No relevant content. **For SPaG marks, please see table in Weather hazards section.**	9+3 SPaG

Physical landscapes in the UK: Coastal landscapes in the UK		
Question	**Marking guidance**	**Total marks**
1	The feature is a spit.	1
2	The spit is formed by the process of longshore drift.	1
3	B 318897	1

4	**Indicative content** • Constructive waves are gently sloping whereas destructive waves are steep. • Constructive waves have a strong swash, which helps to build up beaches whereas destructive waves have a strong backwash, which removes beach material. • Constructive waves have a long wavelength but destructive waves have a short wavelength. **Level 2 (Clear) 3–4 marks:** • AO2 Shows clear understanding of why constructive and destructive waves are different. • AO3 Demonstrates application of knowledge and understanding to make full interpretation of the processes that cause constructive and destructive waves. **Level 1 (Basic) 1–2 marks:** • AO2 Shows limited understanding of the difference between constructive and destructive waves. • AO3 Demonstrates application of knowledge and understanding to make limited interpretations of the processes that cause constructive and destructive waves. • 0 marks: No relevant content.	4
5	**Indicative content** • A large crack forms in the headland at a point of weakness. • The crack is enlarged by the process of hydraulic action and abrasion to form a notch. • Waves make the crack larger over time to form a cave. • Two caves form back-to-back over time. • These will eventually cut through the headland to form an arch. **Level 2 (Clear) 3–4 marks:** • AO2 Shows clear understanding of how an arch is formed. • AO3 Application is sound with clear interpretation of processes involved. **Level 1 (Basic) 1–2 marks:** • AO2 Shows limited understanding of how an arch is formed. • AO3 Application is limited with basic interpretation of processes involved. • 0 marks: No relevant content.	4
6	**Indicative content** • Process of longshore drift transports material along the coast in a zigzag movement, with the swash bringing material up the beach at a 45-degree angle and then the backwash removing it at a 90-degree angle. • When there is a change in direction of the coast, such as at a river estuary or headland, the sea loses energy and material is deposited. • The process is repeated over time and a spit is formed extending out across the estuary. • A change in wind direction causes the spit to create a hooked or curved end. • A salt marsh develops behind the spit. • Other depositional features may also be described – tombolos and bar. These should follow the same ideas about deposition occurring when the sea is calm and has less energy to transport material. **Level 3 (Detailed) 5–6 marks** • AO3 Demonstrates thorough application of knowledge and understanding to give detailed explanations of depositional landforms. • AO3 Shows full understanding of the interrelationships between coastal environments and processes. **Level 2 (Clear) 3–4 marks:** • AO1 Demonstrates specific and accurate knowledge of depositional processes and the landforms they create along the coast. • AO2 Shows clear geographical understanding of the interrelationships between coastal environments and processes. **Level 1 (Basic) 1–2 marks:** • AO1 Demonstrates some knowledge of depositional processes. Alternatively, may be descriptive about landform appearance and structure. • AO2 Shows limited geographical understanding of the interrelationships between coastal environments and processes. • 0 marks: No relevant content.	6

Physical landscapes in the UK: River landscapes in the UK		
Question	**Marking guidance**	**Total marks**
1a	Lag time is the length of time between the peak rainfall and the river's peak discharge.	1
1b	Peak discharge – peak rainfall. 6.8 – 3.5 = 3.3 hours.	1
2	**Indicative content** Physical factors • Precipitation – heavy rainfall can lead to flash flooding or a long period of rain can lead to flooding. • Geology – impermeable rock increases surface run-off, which increases rate of which water reaches a river. • Saturated soil – waterlogged soil doesn't allow water to infiltrate. Human factors • Deforestation – this decreases the amount of interception so rainfall can get to rivers quicker. • Urbanisation – increased impermeable surfaces which increases speed of which water reaches rivers.	4
3	Answers using evidence from the map, including grid references if available: • Source of River Wye shown in grid square 80;87. • Upland area shown by the contour lines being close together. • Peak high ground at 724 m in grid square 81:87. • Shape of contours formed in a v shape with River Wye at the bottom.	4
4	As it runs from source to mouth, the cross-section of a river becomes: • wider • deeper.	2
5	**Indicative content** • There will be larger and more angular sediment in the upper course. • The sediment will be considerably smaller and rounder in the lower course. • Sediment will be rounder and smoother as sediment is transported and eroded as it is moved downstream. • Erosion processes of hydraulic action, attrition and abrasion are key in reducing the size of sediment as it moves downstream. **Level 2 (Clear) 3–4 marks:** • AO2 Shows clear understanding of how and why sediment changes from source to mouth. • AO3 Application is sound with clear interpretation of processes involved. **Level 1 (Basic) 1–2 marks:** • AO2 Shows limited understanding of how size of sediment changes from source to mouth. • AO3 Application is limited with basic interpretation of processes involved. • 0 marks: No relevant content.	4

| 6 | **Indicative content**
• Floodplains are found in the lower stage, of a river.
• Formed when a meander migrated from side to side causing, the valley sides to erode and become wider. This creates a wide valley floor.
• During a period of floods, silt is deposited on the floodplain and builds up layers of fertile land.

Level 2 (Clear) 3–4 marks:
• AO2 Shows clear understanding of how floodplains are formed.
• AO3 Demonstrates application of knowledge and understanding to make a full interpretation of the photograph, suggesting how floodplains are formed.

Level 1 (Basic) 1–2 marks:
• AO2 Shows limited understanding of how flood plains are formed.
• AO3 Demonstrates application of knowledge and understanding to make limited interpretation of photo suggesting possible reasons for the formation of floodplains.
• 0 marks: No relevant content. | 4 |

Physical landscapes in the UK: Glacial landscapes in the UK		
Question	**Marking guidance**	**Total marks**
1	Arête	1
2a	 **Figure 6**	4
2b	One of: • hanging valley • pyramidal peak • corrie.	1
3	**Indicative content** • Drumlins are smooth egg-shaped hills formed by glacial deposition. • Drumlins are made of moraine deposited by a glacier. • They have a streamlined shape created by the ice shaping them as it moved over the moraine to give a blunt end up the valley and a sharp end pointing down the valley. This shows the direction of the glacier's movement. • They can be found on the floor of a glacial trough. **Level 2 (Clear) 3–4 marks:** • AO2 Shows clear understanding of how drumlins are formed. • AO3 Demonstrates application of knowledge and understanding that is sound with clear interpretation of processes involved. **Level 1 (Basic) 1–2 marks:** • AO2 Shows limited understanding of how drumlins are formed. • AO3 Demonstrates application of knowledge and understanding is limited with basic interpretation of processes involved. • 0 marks: No relevant content.	4
4	**Indicative content** Includes any two from a range of economic activities in glacial areas. There is a range of economic activities in glacial areas and the response should include any two of the following activities: • Farming: upland areas have thin acidic soils due to glaciation so are good for sheep farming. • Farming: lowland glacial areas have a thick layer of till making it good for arable farming, e.g. East Anglia. • Forestry: coniferous trees cope well with the acidic soils in upland, e.g. in Scotland. • Quarrying: upland glaciated areas are often made of hard resistant rock. This is good for being crushed and used for stone in the building industry, e.g. limestone from the Pennine Hills. • Tourism: glacial uplands create stunning scenery. This provides excellent tourist opportunities, e.g. the Lake District where tourists spend approximately £1100 million every year. • Tourism: glacial uplands often provide excellent skiing opportunities, providing many jobs, e.g. in hotels, shops and other tertiary services. **Level 2 (Clear) 3–4 marks:** • AO2 Shows clear understanding of how economic activities have developed in upland glaciated areas. • AO3 Demonstrates application of knowledge and understanding that is sound with clear interpretation of economic activities involved. **Level 1 (Basic) 1–2 marks:** • AO2 Shows limited understanding of how economic activities have developed in upland areas. • AO3 Demonstrates application of knowledge and understanding that is limited with basic interpretation of economic activities involved. • 0 marks: No relevant content.	4

Urban issues and challenges: Global patterns of urban change		
Question	**Marking guidance**	**Total marks**
1	One mark for each correct answer: **B** The population of the world experienced a rapid growth from 1960. **C** Global population is predicted to rise to 9 billion by 2050. No credit if three or more answers are shaded.	2
2	• Process 1: Natural increase (1) when the birth rate exceeds the death rate (1). • Process 2: Migration (1) when immigration exceeds emigration in an area (1).	4
3	Two of the following points. • majority of the world's megacities are located in LICs or NEEs • most found in Asia • only five in developed countries • only two in Africa • only one in Europe and none in Oceania • more in the northern hemisphere.	2

4	1950 urban population = approximately 800 million 2050 estimated urban population = approximately 6300 million, so increase is approximately 5500 million or 5.5 billion. Accept range of 5.4 to 5.6 billion.	1
5	**Indicative content** Increase in urban population due to: • rural to urban migration • pull factors encouraging people to move to the cities: better opportunities in urban areas, e.g. jobs, education, health care, infrastructure, housing, etc. • TNCs have increased employment opportunities mostly in the cities • younger people most likely to migrate and are more likely to be most fertile and have a family • rates of urban growth fastest in NEEs/LICs where rural to urban migration is most rapid. Decrease in rural population due to: • push factors forcing people to leave the countryside, e.g. lack of jobs • lack of public services, schools, health care, housing, etc. • lack of money from subsistence farming • drought/crop failure • natural disasters • older people left behind, who are no longer fertile or having a family. **Level 3 (Detailed) 5–6 marks:** • AO3 Demonstrates thorough application of knowledge and understanding to give detailed explanations of reasons for a global increase in population. • AO3 Shows full understanding of both migration and natural increase as reasons for growth in global population in urban areas. Will give some explanation of differences between HICs, NEEs and LICs. **Level 2 (Clear) 3–4 marks:** • AO1 Demonstrates specific and accurate knowledge of reasons for growth in global urban population. • AO2 Shows clear geographical understanding of both migration and natural increase as reasons for growth in global population in urban areas. May recognise the differences between HICs, NEEs and LICs. **Level 1 (Basic) 1–2 marks:** • AO1 Demonstrates some knowledge of reasons for growth in global urban population. • AO2 Shows limited geographical understanding of the different reasons, with a focus on migration or natural increase. • 0 marks: No relevant content.	6

Urban issues and challenges: Urban growth in LICs and NEEs		
Question	**Marking guidance**	**Total marks**
1	Requires two separate points with elaboration. Any two advantages from: • coastal location, with five ports for import and export • centre of culture in Brazil with museums and world-famous carnival • centre for tourism with its beaches and mountainous surroundings, and is a UNESCO World Heritage Site • centre in Brazil for service and manufacturing industries.	4
2	**Indicative content** Informal economy • low paid • no job security • no benefits such as sick leave • no workers' employee rights • no tax paid to the government so doesn't benefit the people in the long term with services such as health and education. **Level 2 (Clear) 3–4 marks:** • AO2 Shows clear understanding of how informal economy creates problems. • AO3 Application is sound with clear interpretation of problems for people in LIC or NEE cities. **Level 1 (Basic) 1–2 marks:** • AO2 Shows limited understanding of how the informal sector economy creates problems. • AO3 Application is limited with basic interpretation of what informal economy is and how it can create problems for people. • 0 marks: No relevant content.	4
3	**Labels may include:** • Open sewers are unhygienic and can attract rats, which spread disease. • Lack of clean running water leads to people drinking contaminated water from the rivers and becoming ill. • Houses (self-)built from wood, plastic sheeting and corrugated iron are unstable and could easily be destroyed in a flood or fire. • Houses are very crowded together with whole families of more than 10 people living in a single room. **Level 2 (Clear) 3–4 marks:** • AO2 Shows clear understanding of how living in a squatter settlement creates problems. • AO3 Application is sound with clear interpretation of evidence in the photograph. **Level 1 (Basic) 1–2 marks:** • AO2 Shows limited understanding of how living in a squatter settlement creates problems. • AO3 Application is limited with basic interpretation of evidence in the photograph. • 0 marks: No relevant content.	4
4	Car	1
5	One mark for each problem identified. One mark for the elaboration of each point. • Problem 1: Pollution, vehicle exhaust fumes, especially diesel engines, cause air pollution. This often creates smog which can lead to health problems such as asthma. • Problem 2: Congestion from volume of traffic and a lack of road awareness creating chaos on the streets.	4

Answers

	Urban issues and challenges: Urban change in the UK	
Question	**Marking guidance**	**Total marks**
1	Include any two points of data from the chart for 1 mark each • Majority of the population are White British • Equal numbers of Black, South Asian and White other	2
2	One mark each for two examples of cultural diversity. One mark each for an elaboration of how this benefits the community. • Areas such as Shoreditch in London have developed into eclectic areas with museums, art galleries, restaurants and pubs, and large amounts of ethnic diversity. • Cultural life has become enriched with a wide range of foods and restaurants from all over the world. • It has developed the economy in the leisure and entertainment industries. • Strong community spirit developed in ethnic groups with events such as the Notting Hill Carnival bringing together black African and Afro-Caribbean people in celebration and creating a major tourist attraction.	4
3	**Indicative content** • Richmond upon Thames has a higher quality of life and a lower level of deprivation than Newham. • On average, people in Richmond upon Thames live to be 83 years old compared to 77 years old in Newham (1). This indicates differences in health and lifestyle between Richmond upon Thames and Newham (1). • A higher percentage of pupils get five A*–C grades in Richmond upon Thames, at 70% compared to 57% in Newham (1) This could be: due to the quality of the schools in Richmond upon Thames (1); due to the differences in attitudes to education of the young people in Newham (1); due to more investment in education and schools in Richmond upon Thames. • People in Richmond upon Thames earn on average nearly twice the salary of people in Newham (1). This could be linked to levels of education, as fewer people in Newham have received higher education (1). More people in Richmond upon Thames are employed in professional positions, having been to university (1). **Level 2 (Clear) 3–4 marks:** • AO2 Shows clear understanding of the differences between two boroughs. • AO3 Application is sound with clear interpretation of evidence in the table. **Level 1 (Basic) 1–2 marks:** • AO2 Shows limited understanding of the differences between two boroughs. • AO3 Application is limited with basic interpretation of evidence in the table. • 0 marks: No relevant content.	4
4	Figure 9 shows the Olympic Park in Stratford. However, the student's answer will depend on the case study used. **Indicative content** Improvements in environmental quality as part of a regeneration project such as Newham included: • derelict and toxic old industrial land was cleaned and could be rebuilt on • waterways and canals cleaned of all pollution • parkland created and 4000 trees planted • new wildlife habitats created. Improvements in social quality as part of a regeneration project such as Newham included: • road and transport improvements • housing improvements. **Level 2 (Clear) 3–4 marks:** • AO2 Shows clear understanding of how the environment has improved as a result of a regeneration project. • AO3 Application is sound with clear interpretation of evidence in the picture and example used. **Level 1 (Basic) 1 mark:** • AO2 Shows limited understanding of how the environment has improved as a result of a regeneration project. • AO3 Application is limited with basic interpretation of evidence in the picture and example used. • 0 marks: No relevant content.	4

	Urban issues and challenges: Urban sustainability in the UK	
Question	**Marking guidance**	**Total marks**
1	**D** Urban and rural areas both consume energy and resources to provide for the population living there. No credit for more than one answer shaded.	1
2	• Inputs such as – food, water, energy, resources, etc. • Outputs such as – water, sewage, pollution, etc.	2
3	Any two features from: • Green spaces (1) – creation of parkland to improve environmental quality (1). • Affordable housing (1) – with low rents to make housing affordable for local people (1). • Energy efficient housing (1) – with high levels of insulation, triple glazing and energy efficient appliances to reduce energy consumption and heat loss (1). • Transport provision (1) – bus routes, cycle lanes, car shares schemes to reduce reliance on individual cars (1). • Local services and facilities provided within community (1) – schools, health centre, and community centre (1).	4
4	**Indicative content** • Doesn't create any harmful CO_2, which causes global warming. • Improves pollution and smog levels in cities, which improves the health of people breathing in fumes. • Provides a form of fitness for the people. • Is a sociable activity to do with other people. • Encourages people to get out of their cars if safe cycle lanes are available. • Can be used as a way to reach an integrated transport system such as a train, tube or bus station. **Level 2 (Clear) 3–4 marks:** • AO2 Shows clear understanding of sustainable transport strategies. • AO3 Application is sound with clear interpretation of how cycling can be an important part of a sustainable transport strategy. **Level 1 (Basic) 1 mark:** • AO2 Shows limited understanding of sustainable transport strategies. • AO3 Application is limited with basic interpretation of how cycling is part of a sustainable transport strategy. • 0 marks: No relevant content.	4

5	The student needs to evaluate a scheme using a named case study, e.g. London or Bristol, and explain the strategies used.	6
	Indicative content This will vary depending on the example chosen but could include: • integrated transport network linking all different transport methods together • use of ticket systems like the Oyster card to make travel easier and cheaper • developing use of cycle lanes to encourage more cycling • use of bus lanes to speed up bus travel • more frequent public transport services • improvements to tube services and trains in London • sustainable bus services – electric buses, Bristol's Poo Bus • electric car charging points • congestion charge in London • ring roads to take traffic around city centre • park-and-ride schemes to avoid going into city centre Problems or failures of the schemes: • cycle lanes can't be created on narrow roads • cycle lanes have parked cars or trees in them so not useable • crowding on public transport • people still like the convenience of their cars. **Level 3 (Detailed) 5–6 marks:** • AO3 Demonstrates thorough application of knowledge and understanding to give detailed evaluation of the strategies to manage traffic congestion. • AO3 Shows full understanding of a variety of schemes and gives named examples. Clearly shows the benefits and problems of such schemes. • Offers a conclusion. **Level 2 (Clear) 3–4 marks:** • AO2 Demonstrates specific and accurate knowledge of two or more different strategies to manage traffic congestion. • AO3 Shows clear geographical understanding to evaluate the success of schemes using at least one specific example. **Level 1 (Basic) 1 mark:** • AO1 Demonstrates some knowledge of strategies to manage traffic congestion. • AO2 Shows limited geographical understanding of how successful different schemes have been and may not give a suitable example. • 0 marks: No relevant content.	
6	Two advantages from: • Improve air quality (1), which benefits people as air is cleaner for breathing, improving health (1). • Creates wildlife habitats (1) and regenerates species in urban areas (1). • Creates social spaces (1) for people to improve the community and the social sustainability of an area. (1)	4

The changing economic world: Economic development and quality of life

Question	Marking guidance	Total marks
1	Credit similarities and differences between South America and Africa. **Two** should be described or implied, such as: • Africa has many more LIC countries than South America (1) • South America only has two LICs while Africa has many (1) • Neither continent has any HICs (1) Credit relevant reference to named countries. No credit for descriptions of other continents or global pattern. Answer must also offer a comparison.	2
2	Credit **one** advantage only. One mark for basic statement, such as: • It shows how wealthy people are (1). • It shows what people are getting paid for their work (1). Two marks for developed idea, such as: • It shows the wealth of people and to what extent they can improve their lives by spending money on their homes or services (2). • It shows how much money people are earning and therefore how much taxes the government could be getting to spend on improving infrastructure (2).	2
3	Credit **one** disadvantage only. One mark for basic statement, such as: • It only considers social and economic indicators (1). • It does not use equality (1). Two marks for developed idea, such as: • It does not take into account the state of the natural environment, which if damaged would make development unsustainable (2). • The social factors that it includes do not consider gender inequalities or other inequalities such as wealth and ethnicity (2).	2
4	**Indicative content** • Climate influences development over a long period of time, with greater development in cooler wetter areas (temperate zones). • Where it is not too hot and not too cold with enough rain, crops grow better, so there is more food for people. • There are fewer fatal diseases in cooler areas, as these thrive where it is hot, like the tropics, where many people die or are weakened by disease so they cannot work well. • Hot and cold environments and drier climates are more difficult to work in. The human body cannot operate in extremes and therefore work is less efficient. **Level 2 (Clear) 3–4 marks:** • AO2 Shows sound understanding of one aspect of physical geography relevant to development. • AO3 Demonstrates sound application of knowledge and understanding in explaining how the chosen aspect of physical geography has affected development. **Level 1 (Basic) 1–2 marks:** • AO2 Shows limited understanding of one aspect of physical geography relevant to development. • AO3 Demonstrates limited application of knowledge and understanding in explaining how the chosen aspect of physical geography has affected development. • 0 marks: No relevant content.	4
5	Haiti = 200 per 100 000 people. USA = 3 per 100 000 people. So 200 − 3 = **197** per 100 000 people.	1

Answers

6	**Indicative content** • The answer must give reasons for differences; there is no credit available for description of the graph on its own. • The graph shows that HICs have the lowest TB cases and LICs have the highest, except for South Africa which has the highest level of TB. • HICs have a lot of money and have invested in health care systems, medicines and doctors, which keep the population healthier. • LICs are poor and do not have the money to invest in medical care and so people catch diseases more easily and are often not treated. • LICs are poor and people cannot afford medical care and treatment. • In LICs people live in poor conditions with lower levels of hygiene which allows TB to exist and spread. • South Africa may classify as a NEE economically but it is a country that has not invested in social infrastructure to improve all people's lives at work and at home. • LICs and NEEs may not be able to help people in remoter rural areas where TB rates can be higher. **Level 2 (Clear) 3–4 marks:** • AO2 Shows sound understanding of reasons for differences in cases of TB. • AO3 Demonstrates sound application of knowledge and understanding in explaining why the cases of TB vary. **Level 1 (Basic) 1–2 marks:** • AO2 Shows limited understanding of reasons for differences in cases of TB. • AO3 Demonstrates limited application of knowledge and understanding in explaining why the cases of TB vary. • 0 marks: No relevant content.	4
7	**Two** reasons should be given, such as: • People are not strong enough to work efficiently on farms; this creates less food supplies, which weakens people further. • People are not strong enough to work efficiently in factories, so productivity goes down and less money is made. • Children may not develop properly and cannot contribute to the economy as adults. • If people are unable to work properly and earn money they cannot pay taxes, so the government has less money to invest in infrastructure and services. No credit for vague statements such as: too weak to work; health is poor; low-paid jobs.	2

The changing economic world: Global development gap		
Question	**Marking guidance**	**Total marks**
1a	Responses must focus on the comparison. Statement may be backed by fact. • FDI outflows have increased greatly from Developing Asia while from Europe they have declined. • Developing Asia increased by \$133b while Europe declined by \$60b.	1
1b	Transition = \$63b and Africa = \$13b. 63 − 13 = **\$50b**	1
2	Credit **one** way only. One mark for basic statement, such as: • FDI provides jobs (1) • FDI helps the multiplier effect (1). Two marks for developed idea, such as: • FDI brings in investment and creates jobs and therefore creates income for the country (2). • FDI may bring factories into the country which will encourage the growth of linked industries, e.g. suppliers and services (2).	2
3	**One** expanded reason should be given, such as: • FDI may not reduce poverty because the amount of investment is too small to have a large impact, such as only employing a few people on low wages. • FDI may not reduce poverty because the TNC may take the profits made out of the country and not re-invest it to create economic growth locally. • FDI may not reduce poverty because the TNC uses components and supplies from abroad and so does not support local businesses and industries. No credit for vague statements such as: • doesn't make people richer; profits disappear • government gives too much money.	2
4a	2015 tourists = 3.7 m tourist arrivals and 2009 tourists = 2.75m. 3.7 − 2.75 = **0.95m** (allow 0.9m to 1.0m)	1
4b	Responses must focus on the comparison. Statement may be backed by fact. Only one response required. • Tourist arrivals increased steadily between 2009 and 2015 with minor fluctuations, while the income fell between 2009 and 2011 before increasing slowly after 2012. • Tourist income was highest in 2009 when arrivals were at their lowest, then as arrivals increased, income decreased until after 2012, the gap between arrivals and income has been increasing.	1
5	Credit **one** way only. One mark for basic statement, such as: • Tourism provides jobs for people (1). • Tourism money helps start the multiplier effect (1). Two marks for developed idea, such as: • Tourism brings money into the country, which gives the government tax money that it can spend on improving services such as education. (2). • Tourism provides many jobs and so gives people wages, these people then spend money in local shops and for local services which gives more people wages, which they can then spend on other things. In this way the cycle of poverty is broken (2).	2
6	One disadvantage for one mark, such as: • tourism may provide only seasonal employment (1). **One** expanded disadvantage should be given for two marks, such as: • Tourism may be a disadvantage because the country may not receive all of the money as the tourism TNCs take most of it, and the amount of money received can vary each year so income is unreliable (2). • Tourism may be a disadvantage because large numbers of people put pressures on infrastructure and the natural environment; if infrastructure cannot cope then pollution and damage to the natural environment may occur (2). • Tourism may be focused in one part of a country and other areas may not receive the benefits of tourist income at all, creating a wealth and quality of life division within the country (2). No credit for vague statements such as: doesn't benefit everyone; income is low; government favours tourist areas.	2

7	The question asks for justification as to whether or not there have been significant improvements to the quality of life of poor people in an LIC or NEE brought by industrial change. It requires agreement or disagreement with the statement backed by evidence. **Indicative content** • Economic and industrial development from agriculture to industry produces more money, which can be invested in improving infrastructure such as water and sanitation: these help to improve the health of people. • Secondary industries are able to provide better paid jobs than farming and so the wages of people increase beyond the poverty level of US$1.90 per day; with more money people are able to improve their living conditions – such as their homes. • Economic and industrial development may increase trade and external links but the LIC/NEE may not benefit, ending up with a negative trade balance so that the country gets into debt and has no money to improve infrastructure that would improve people's lives. • Industrial development may occur too fast so that environmental laws are not in place; this may lead to pollution of the natural environment which may affect the living conditions (e.g. the Ogoni in Nigeria affected by oil spills). **Level 3 (Detailed) 7–9 marks:** • AO1 Demonstrates comprehensive and specific knowledge of industrial change and quality of life. • AO2 Shows thorough and accurate geographical understanding of industrial change and quality of life. • AO3 Demonstrates effective application of knowledge and understanding in making a judgement about the issues and reaching a substantiated conclusion. Justification is detailed and balanced. **Level 2 (Clear) 4–6 marks:** • AO1 Demonstrates reasonable knowledge of industrial change and quality of life. • AO2 Shows some geographical understanding of industrial change and quality of life. • AO3 Includes reasonable application of knowledge and understanding in making a judgement about the issues and reaching a conclusion. Justification is clear and well supported. **Level 1 (Basic) 1–3 marks:** • AO1 Demonstrates limited knowledge of industrial change and quality of life. • AO2 Shows some geographical understanding of industrial change and quality of life. • AO3 May either include limited application of knowledge and understanding in making a judgement about the issues or in reaching a conclusion. Justification is limited to one or more simple points. • 0 marks: No relevant content and judgement not made. **For SPaG marks, please see the table in Weather hazards above.**	9+3 SPaG

The changing economic world: Rapid economic development and change		
Question	**Marking guidance**	**Total marks**
1a	2004 = 2350 thousand of barrels per day, 2016 = 1750 thousand of barrels per day. 2350 − 1750 = **600** thousand barrels per day. (Allow 575 to 625.)	1
1b	Response must focus on the comparison. Statement may be backed by fact. Only **one** response required. • Between 2004 and 2007 oil production increased to a peak and then declined sharply to a lower level, while between 2007 and 2010 there was a steep decline followed by a rise. • Both time periods have fluctuations but most of the 2004 to 2007 time period is at a high level (over 2400 thousand barrels per day) while 2007 to 2010 has a sharp decline – even dipping below 1700 thousand barrels per day.	1
2	Credit **one** disadvantage only. Can be social or political. Should be linked to a named LIC/NEE country. Nigeria is used as an example in the RG. One mark for basic statement, such as: • Nigeria has ethnic divisions (1) • Nigeria has corrupt politicians (1). Two marks for developed idea, such as: • Nigeria has three main ethnic groups and several minor ones. Historically there has been social division between the main groups which has led to internal conflicts and held back improvements to the quality of life (2). • Nigeria has a lot of corruption at decision-making levels within the country; some of this is linked to oil wealth. Therefore decisions have not led to efficient use of the oil wealth and quality of life in many areas remains poor (2).	2
3	**Indicative content** • Nigeria had the advantages of grasslands for growing crops and rearing animals in the centre of the country and agroforestry/plantations in the southern tropical rainforests. However, income from these was relatively low when exported. • Nigeria has changed from a country largely dependent on agricultural products to an oil producer with recent expansion of secondary industries. • The discovery of oil offshore led to Nigeria becoming the world's 12th largest producer and brought in a lot of wealth from exports (oil makes up 91% of exports). • Oil is a finite resource and will run out; therefore some of the wealth has been invested in secondary industries, e.g. soap manufacturing, to provide jobs and alternative income for the future. Some TNCs have invested in the country (e.g. Unilever). **Level 2 (Clear) 3–4 marks:** • AO2 Shows sound understanding of the changing industrial structure of an LIC/NEE. • AO3 Demonstrates sound application of knowledge and understanding in explaining why the LIC/NEE industrial structure has changed. **Level 1 (Basic) 1–2 marks:** • AO2 Shows limited understanding of the changing industrial structure of an LIC/NEE. • AO3 Demonstrates limited application of knowledge and understanding in explaining why the LIC/NEE industrial structure has changed. • 0 marks: No relevant content	4

Answers

<table>
<tr>
<td>4</td>
<td>
The question asks for justification as to whether or not international aid and trade are essential to the economic development of LIC or NEE countries. It requires agreement or disagreement with the statement, backed by evidence.

Indicative content
- Economic development may be brought about by several things, international aid and trade are just two of these, others include remittances from migrants, development of infrastructure, tourism, introducing intermediate technology, getting rid of debts, and encouraging microfinance loans.
- International aid may take several forms; loans may be provided to help build expensive development schemes, or may be smaller scale from NGOs for bottom-up local schemes to improve the lives of people. Nigeria receives aid for social issues such as clean water, education and health care as it has large numbers of poor people and cannot help them all.
- Trade brings in money to a country by selling goods and resources that other countries want to buy. For example, Nigeria has oil, which is in great demand, and so the country can make considerable amounts of money (supplying 2.7% of the world's oil); it is expected to become a world top 20 economic nation soon because of this.
- However, international aid may create dependency on external help and large loans may lead to debt. Trade is also not always fair, especially where raw materials are being sold which fetch a lower price than processed goods (e.g. turning oil in petrol). In 2016, Nigeria had a negative trade balance.

Level 3 (Detailed) 7–9 marks:
- AO1 Demonstrates comprehensive and specific knowledge of international aid and trade of an LIC/NEE.
- AO2 Shows thorough and accurate geographical understanding of international aid and trade of an LIC/NEE.
- AO3 Demonstrates effective application of knowledge and understanding in making a judgement about the issues and reaching a substantiated conclusion. Justification is detailed and balanced.

Level 2 (Clear) 4–6 marks:
- AO1 Demonstrates reasonable knowledge of international aid and trade of an LIC/NEE.
- AO2 Shows some geographical understanding of international aid and trade of an LIC/NEE.
- AO3 Includes reasonable application of knowledge and understanding in making a judgement about the issues and reaching a conclusion. Justification is clear and well supported.

Level 1 (Basic) 1–3 marks:
- AO1 Demonstrates limited knowledge of international aid and trade of an LIC/NEE.
- AO2 Shows some geographical understanding of international aid and trade of an LIC/NEE.
- AO3 May either include limited application of knowledge and understanding in making a judgement about the issues or in reaching a conclusion. Justification is limited to one or more simple points.
- 0 marks: No relevant content and judgement not made.

For SPaG marks, please see the table in Weather hazards above.
</td>
<td>9+3SPaG</td>
</tr>
</table>

	The changing economic world: Changes in the UK economy	
Question	**Marking guidance**	**Total marks**
1a	Credit similarities and differences between primary employment and tertiary employment. **Two** should be described or implied. Such as: • primary declines from high of 70% to about 10% (1) • tertiary increases from 10% to over 50% (1) • the changes show opposite trends (1). No credit for descriptions of other lines. Answer must also offer a comparison.	2
1b	**Two** reasons should be given, such as: • raw materials (resources) in the UK in short supply/running out so could not supply secondary industries • outdated machinery and inefficiency in industry meant that secondary industries could not compete with more modern industries abroad • competition from Germany, Japan and South Korea, etc., reduced export earnings and made UK industries less profitable causing closures. No credit for vague statements such as industry was old; Japan developed its industries; jobs lost.	2
2	**Indicative content** • The UK now has many tertiary businesses and growing quaternary sector, these depend on the use of information technology to maintain links with each other and worldwide. • London is a world financial centre. This has been possible due to the internet and mobile communications technologies, which enable links to be maintained throughout the world to customers and financial markets. Two million UK people are employed in the financial sector. • 1.3 million UK people work directly in the IT sector of the economy. • Businesses are free to locate anywhere in the UK as office work can just require access to the internet and computers, and people can even work from home. **Level 2 (Clear) 3–4 marks:** • AO2 Shows sound understanding of the role of information technology in post-industrial UK. • AO3 Demonstrates sound application of knowledge and understanding in explaining how information technology creates economic opportunities in the UK. **Level 1 (Basic) 1–2 marks:** • AO2 Shows limited understanding of the role of information technology in post-industrial UK. • AO3 Demonstrates limited application of knowledge and understanding in explaining how information technology creates economic opportunities in the UK. • 0 marks: No relevant content.	4

3	The question asks for agreement as to whether or not it is possible to have economic growth in the UK without significant damage to the physical environment. It requires agreement or disagreement with the statement, backed by evidence. **Indicative content** • Economic growth may bring more factories and greater demand for raw materials such as coal, rock, sand, gravel, oil and metals. All of these activities have potential to damage the land, water, air and ecosystems through emissions or extraction activities. • Power stations provide electricity and these may emit greenhouse gases or radioactive waste or spoil scenery, depending on the primary energy source. As economic growth takes place there is an increase in energy demand which increases the impacts. • Quarrying creates scars in the landscape, such as at Torr Quarry in Somerset. Natural vegetation has also been removed and run-off patterns changed, and there is dust and noise pollution. Due to demand this quarry is predicted to operate until 2040. • The damage from quarries can be repaired; at Torr Quarry the plans are to create new woodland and grassland areas and landscape the quarry to create natural rock features and a lake. • With stronger environmental laws, eco-friendly technologies and greater awareness of potential damage, the physical environment can be protected and damage reduced to below a significant level. A lot of economic growth is in the tertiary and quaternary sectors which require fewer raw materials, just electricity. **Level 3 (Detailed) 7–9 marks:** • AO1 Demonstrates comprehensive and specific knowledge of economic growth and damage to the natural/physical environment in the UK. • AO2 Shows thorough and accurate geographical understanding of economic growth and damage to the environment. • AO3 Demonstrates effective application of knowledge and understanding in making a judgement about the issues and reaching a substantiated conclusion. Justification is detailed and balanced. **Level 2 (Clear) 4–6 marks:** • AO1 Demonstrates reasonable knowledge of economic growth and damage to the natural/physical environment in the UK. • AO2 Shows some geographical understanding of economic growth and damage to the environment. • AO3 Includes reasonable application of knowledge and understanding in making a judgement about the issues and reaching a conclusion. Justification is clear and well supported. **Level 1 (Basic) 1–3 marks:** • AO1 Demonstrates limited knowledge of economic growth and damage to the natural/physical environment in the UK. • AO2 Shows some geographical understanding of economic growth and damage to the environment. • AO3 May either include limited application of knowledge and understanding in making a judgement about the issues or in reaching a conclusion. Justification is limited to one or more simple points. • 0 marks: No relevant content and judgement not made. **For SPaG marks, please see the table in Weather hazards above.**	9+3 SPaG

The challenge of resource management: Resource management		
Question	**Marking guidance**	**Total marks**
1	One valid reason should be given, such as: • New Zealand is on a plate boundary so magma is near the surface to provide heat • New Zealand is an HIC and therefore has the money and technology available to develop geothermal energy. No credit for vague statements such as: • has volcanoes • is a rich country.	1
2	Must be a description of distribution. Credit observations based on the map provided. One mark for basic description, such as: • more in northern hemisphere • most in areas with volcanoes. Second mark for developed point using detail from the map, such as: • Northern hemisphere areas such as USA, Iceland, Italy and Japan have a lot of geothermal energy, while South America and most of Africa have little. • Uses data from the map, e.g. the Asia Pacific region has the greatest amount of geothermal energy capacity installed at 4.81 GW. • Countries and regions with destructive plate boundaries such as New Zealand and South-East Asia have geothermal energy. • Countries and regions with constructive plate boundaries such as Iceland and Kenya have geothermal energy.	2
3	**Indicative content** • Candidates should apply their knowledge and understanding of the energy debate to emphasise the role of sustainable energy supplies. Use should be made of the Figure. • IceLink involves undersea electricity cables directly from Iceland to the rest of Europe, e.g. UK, Norway and Germany. Iceland produces electricity by using geothermal energy (over 500 MW installed) – a renewable and sustainable source. • The energy mix of most European countries, such as the UK, only includes a small proportion of renewable energy sources and still is heavily dependent on fossil fuels. There needs to be a greater diversification into renewables for the future for the benefit of the natural environment but also to achieve energy security. • Countries like Norway and the UK have depended on North Sea oil and natural gas in the past and energy diversification is needed. Most northerly European countries have more limited opportunities for solar energy, but do use wind energy. A geothermal energy source would be an important addition to the future energy mix. • Energy demands have been increasing in countries such as Germany due to growth of tertiary industries and consumers as more electrical gadgets are being used. • Imports of energy from Iceland would avoid some of the negative issues associated with developing renewable energy sources in the UK, such as bioenergy and solar farms using up cropland, and wind turbines spoiling the scenery and causing deaths of birds. **Level 3 (Detailed) 5–6 marks:** • AO3 Demonstrates thorough application of knowledge and understanding of the energy mix of Europe (HICs) or selected European countries. • AO3 Demonstrates thorough application of knowledge and understanding of sustainable energy supplies and an evaluation of why they are needed. **Level 2 (Clear) 3–4 marks):** • AO3 Demonstrates application of knowledge and understanding by making relevant points about the energy mix of Europe (HICs) or selected European countries. • AO2 Develops one or more points that are relevant to sustainable energy supplies and an awareness of its importance. **Level 1 (Basic) 1–2 marks:** • AO3 Demonstrates application of knowledge and understanding through a limited understanding of the energy mix of Europe. • AO2 Demonstrates limited development of one or two points about sustainable energy supplies. • 0 marks: No relevant content.	6
4	One mark for correctly drawn bar. Shading not essential to gain mark.	1
5	One mark for the correct answer. **D** The Solway/Tweed has the best water quality No credit if two or more answers are shaded.	1

Answers

6	Responses should make use of the figure in order to explain their reasons. Candidates should apply their knowledge and understanding in interpreting the resource. One mark for recognition of highland areas with greater rainfall, drier South-East, location of London and demand for water, i.e. source and demand areas. One mark for reasons that focus on the fact that the areas with water sources such as Kielder reservoir and the Lake District are some distance away from the major cities such as Manchester, Leeds, Birmingham and London where water demand is highest. Need a way of transferring water such as through a canal. One mark for the need to irrigate crops in the South-East. No credit for simply describing the map.	3

The challenge of resource management: Food		
Question	**Marking guidance**	**Total marks**
1	One mark for correct answer. **C** 2390–2620 calories No credit if two or more answers are shaded.	1
2	The focus of the question is on where the distribution of daily calorie intake per capita is high and where it is lower in South America. One mark for basic description from the map, such as: • there are some countries with high and some with very low • countries in the east and south are higher than those on the north, west and centre. Second mark for developing a point with detail from the map, such as: • There is one country (Brazil) with 3050–3270 calories in the east and two with 2850–3050 calories in the south (Argentina, Chile), while two countries are below 2170 calories – one in the centre (Bolivia). • The countries in the north and north-west have lower daily calories, such as three with 2390–2620 calories than those in south and east which have between 230 and 650 more calories per person per day. Names of countries not expected. No credit if wrong continent used or just listing of countries.	2
3	Two natural factors should be stated. These can be any physical/environmental factors on a large scale (for a country), such as: • climate (temperature, precipitation, sunlight, seasons) • water supply • rock type and soils • pests and diseases.	2
4	Two reasons should be given, such as: • World population has grown and so there are more people to feed. • People are getting wealthier, even in LICs and especially in NEEs, and therefore people are able to buy more food. • Food tastes change as wealth increases and so there is a higher demand for certain food products.	2
5	**Indicative content** • An understanding of food insecurity and its impacts should be indicated in the answer. Impacts include famine, undernutrition, soil erosion, rising prices, social unrest. • Responses should focus on the difficulties caused for poorer people by a selection of impacts. • Expect some development of at least one situation where there have been food insecurity issues for poorer people. • Famine means that people do not have enough food to live, and the poorest, such as subsistence farmers, will be affected first. LICs in semi-arid areas such as the Sahel are often affected (e.g. Somalia 2010–2012) and people die. • When there are food shortages, prices in markets go up and poorer people cannot afford to buy enough food for a balanced diet; people then become unhealthy and are too weak to fight off disease. Children also become stunted and their development restricted. • The difficulties may become long term if soils are eroded by overuse or poor land management, or it could cause social unrest and armed conflicts as groups try to secure access to food resources (e.g. South Sudan). **Level 3 (Detailed) 5–6 marks:** • AO2 Shows thorough understanding of the impacts caused by food insecurity. • AO2 Demonstrates in detail how these impacts cause difficulties for poorer people. **Level 2 (Clear) 3–4 marks:** • AO1 Demonstrates specific and accurate knowledge of food insecurity, perhaps in a selected location. • AO2 Shows sound understanding of the impacts caused by food insecurity. **Level 1 (Basic) 1–2 marks:** • AO1 Demonstrates limited knowledge of food insecurity, perhaps in a selected location. • AO2 Shows simple understanding of the impacts caused by food insecurity. • 0 marks: No relevant content.	6

The challenge of resource management: Water		
Question	**Marking guidance**	**Total marks**
1	One mark for correct answer. **D** Extremely high stress No credit if two or more answers are shaded.	1
2	The focus of the question is on distribution of water stress in China; where it is high and where it is lower. One mark for basic description from the map, such as: • water stress is extremely high in some places and does not exist in other places • water stress in parts of the north-east and east is higher than in the west and north. Second mark for developing a point with detail from the map, such as: • In the north-east the river basins of the Yongding He, Tuhai Hu, Liao He and Daliao have extremely high water stress levels and the water stress level in Huang He is high. • About half the country of China suffers from some level of water stress; with a quarter of the country, located in the central-east area, having low to medium stress and another quarter medium stress or above. This leaves the rest of the country, especially the west and coastal areas, with no water stress. No credit for wrong country or just listing of river basins.	2
3	Two reasons should be given, such as: • world population has grown and so more water is needed – water is essential for life • people are getting wealthier, even in LICs and especially in NEEs, and therefore people are demanding more water for different purposes • farming has expanded due to more food demand and so more water is needed for irrigation. Farming is the biggest user of water in the world.	2
4	Two reasons should be given, such as: • Rain soaks into the ground and this can be extracted through wells. However, if too much is taken out long-term supplies are reduced, so it is better to restrict how much is taken out. • In times of lower rainfall, the groundwater is not replenished and so supplies can run low; it is therefore necessary to control how much water is taken out of the ground so that there is always enough there for people in the future. • Recharging aquifers, such as in Arizona, can replace water that has been used to be ready for future periods of low rainfall and drought.	2

| 5 | **Indicative content**
• An understanding of water insecurity and its impacts should be indicated in the answer. Impacts include disease, pollution, food production, industrial output, disputes/conflict and social unrest.
• Responses should focus on the difficulties caused for poorer people by a selection of impacts. Nearly 800 million people worldwide do not have access to safe water.
• Expect some development of at least one situation where there have been water insecurity issues for poorer people.
• Clean safe water may not exist for poorer people to use; their health is therefore affected by disease (e.g. cholera such as in Haiti after the earthquake and hurricane) and pollution (e.g. phosphates from farming).
• If there is not enough water, then fields cannot be irrigated and crop yields will be lower or fail in a drought situation. When there are food shortages, prices in markets go up and poorer people cannot afford to buy enough food for a balanced diet, people then become unhealthy and are too weak to fight off disease. Children may also become stunted and their development restricted.
• Factories may be forced to close if they do not have water for cooling machinery or processing; people lose their jobs and have no money.
• Water is essential for life and people may fight over it, especially if a government tries to use it for other purposes so that the poor people do not get as much (e.g. upstream dam).

Level 3 (Detailed) 5–6 marks:
• AO2 Shows thorough understanding of the impacts caused by water insecurity.
• AO2 Demonstrates in detail how these impacts cause difficulties for poorer people.

Level 2 (Clear) 3–4 marks:
• AO1 Demonstrates specific and accurate knowledge of water insecurity, perhaps in a selected location.
• AO2 Shows sound understanding of the impacts caused by water insecurity.

Level 1 (Basic) 1–2 marks:
• AO1 Demonstrates limited knowledge of water insecurity, perhaps in a selected location.
• AO2 Shows simple understanding of the impacts caused by water insecurity.
• 0 marks: No relevant content. | 6 |

The challenge of resource management: Energy		
Question	**Marking guidance**	**Total marks**
1	One mark for correct answer. **C** 40 = 59% No credit if two or more answers are shaded.	1
2	The focus of the question is on where the distribution of energy access and security is high and where it is lower in South America. One mark for basic description from the map, such as: • there are some countries with high and some with low performance levels • there is a random distribution pattern from high to quite low. Second mark for developing a point with detail from the map, such as: • There is one country (Colombia) in the high category over 80%, and three countries in each of the other categories apart from the lowest (0–19%) where there are none. • The countries are split evenly between the categories with no distinct geographical pattern. There are three between 60–79% (e.g. Brazil), three between 40–59% (e.g. Peru), and three between 20–39% (e.g. Venezuela). Only one country in the high category. Names of countries not expected. No credit if wrong continent used or just listing of countries.	2
3	Two reasons should be given, such as: • Insulating walls and roofs reduces energy demand for heating by preventing heat escaping, leaving more energy for the future. • Energy efficient appliances use less electricity so not as much needs to be generated by power stations so they use less fuel, saving more for the future. • People can fit solar panels to their roofs, which means that they can generate some of their own energy from a free renewable resource instead of using fossil fuels such as natural gas.	2
4	**Indicative content** • An understanding of energy insecurity and its impacts should be indicated in the answer. Impacts include: removing people, environmental damage, food production, industrial output, disputes/conflict and social unrest. • Responses should focus on the difficulties caused for poorer people by a selection of impacts. • Expect some development of at least one situation where there have been energy insecurity issues for poorer people. • As energy runs out, there are pressures to expand exploration, extraction and production. Some of the locations chosen for this impact on poorer people, such as the Inuit in the Arctic or the Nanti in Peru. Poorer people may be forced to move from their homes because of flooding behind an HEP dam, or pipeline construction. • When there are energy shortages prices go up and poorer people cannot afford to buy enough energy for their needs. This affects the quality of their life and small businesses such as workshops and crafts. • Factories may be forced to close if they do not have energy to operate; people lose their jobs and have no money. • Energy is regarded by many as a modern essential and there is a lot of money involved so people may fight over it; often it is the poor people who are caught in the middle of conflicts (e.g. Iraq). **Level 3 (Detailed) 5–6 marks:** • AO2 Shows thorough understanding of the impacts caused by energy insecurity. • AO2 Demonstrates in detail how these impacts cause difficulties for poorer people. **Level 2 (Clear) 3–4 marks:** • AO1 Demonstrates specific and accurate knowledge of energy insecurity, perhaps in a selected location. • AO2 Shows sound understanding of the impacts caused by energy insecurity. **Level 1 (Basic) 1–2 marks:** • AO1 Demonstrates limited knowledge of energy insecurity, perhaps in a selected location. • AO2 Shows simple understanding of the impacts caused by energy insecurity. • 0 marks: No relevant content.	6
5	Two reasons should be given, such as: • world population has grown and so therefore more energy is needed (e.g. for cooking, heating) • people are getting wealthier, even in LICs and especially in NEEs, and therefore people are demanding more energy for different purposes in their homes and lives (e.g. appliances, gadgets, cars) • industries and businesses are expanding as the world develops economically and all of these need energy for factories, offices and transport.	2

Answers to the practice papers are available online.
Visit: www.scholastic.co.uk/gcse

Exam Practice acknowledgements

Revision Guide acknowledgements

Photo and artwork permissions

p.18 Jamilia Marini/Shutterstock.com; p.19 (top) Image by Robert A. Rohde, Global Warming Art, Creative Commons Attribution-Share Alike 3.0 Unported license; p.21 Track of Typhoon Haiyan, redrawn from www.mapsoftheworld.com; p. 25 © Jonathan Mitchell/Dreamstime.com; p.27 daulon/Shutterstock.com; p.34 (top) Frontpage/Shutterstock.com; p.34 (bottom) Redrawn from Soil climate map, USDA-NRCS, Soil Science Division, World Soil Resources, Washington DC; p.37 NASA Goddard Space Flight Center Image by Reto Stöckli (land surface, shallow water, clouds). Enhancements by Robert Simmon (ocean color, compositing, 3D globes, animation). Data and technical support: MODIS Land Group; MODIS Science Data Support Team; MODIS Atmosphere Group; MODIS Ocean Group Additional data: USGS EROS Data Center (topography); USGS Terrestrial Remote Sensing Flagstaff Field Center (Antarctica); Defense Meteorological Satellite Program (city lights); p.41 Redrawn from map by UN Environment, sources: Brazilian Institute for Geography and Statistics; La déforestation en Amazonie; p.43 Global Warming Images / Alamy Stock Photo; p.49 George Steinmetz; p. 51 Adapted from Great Green Wall Initiative; p.56 Wikimedia/Hannes Grobe, Alfred Wegener Institute, Creative Commons Attribution SA 2.5; p.57 Pete Bucktrout, British Antarctic Survey; p.60 Bardocz Peter/Shutterstock.com; p. 63 Adapted from image by yefi/Wikimedia Commons; p. 64 Adapted from image at http://www.bbc.co.uk/education/guides/z3ndmp3/revision/3; p.65 Adapted from diagram by Anna Bozzo, https://102coastsgroup2.wikispaces.com, Creative Commons SA 3.0 licence; p. 65 (top) Adapted from http://thebritishgeographer.weebly.com/uploads/1/1/8/1/11812015/2239101.gif?875; p. 66 (bottom) Adapted from diagram at https://www.geocaching.com/geocache/GC4KF9W_gold-coast-seaway-who-gives-a-spit?guid=d6d89c9a-de21-4fc6-a23b-7d374bf245f1; p.72 (top) Dave Head/Shutterstock.com; p.78 (top) Chris Lofty | Dreamstime.com; p.80 Galyna Andrushko/Shutterstock.com; p.82 (bottom) Drew Rawcliffe/Shutterstock.com; p. 88 Clevercapybara/Wikimedia Commons; p. 89 Data source UN 2002; p.91: Skreidzeleu/Shutterstock.com; p. 97 Source: ONS; p.102 (top) Alexandre Rotenberg/Shutterstock.com; p.102 (bottom) Angelina Dimitrova/Shutterstock.com; p.104 chrisdorney/Shutterstock.com; p.109 World Bank Databank 2016; p.110 Redrawn from http://hdr.undp.org/en/countries; p. 111 Source World Bank; p. 115 Source World Bank; p.119 (top) Fairtrade Foundation; p. 119 (bottom) Redrawn from Microfinance Market Outlook, 2016, responsibility; p. 121 Source Jamaica Tourist Board; p. 123 (top) Adapted from Miniwatts Marketing Group. Data: Internet World Stats; p.123 (bottom) www.tradingeconomics.com, Organisation of the Petroleum Exporting Countries; p.124 REUTERS / Alamy Stock Photo; p. 125 Data from Trading Economics; p.131: © Crown copyright 2017 OS [licence number pending]; p.133 Redrawn from image by Tom Forth/CityMetric; p.134 © Mr LG Frost; p.135 Gatwick Airport Limited; p. 136 Heathrow Airport Holdings Limited; p. 137 Data © crown copyright and database rights (March 2014) Ordnance Survey; p. 142 FAO June 2016 Prospects and Situation; p. 143 Source FAO; p.145 (top) RWTH Aachen University, Department of Engineering Geology and Hydrogeology; p.146 Adapted from *BP Statistical Review of World Engergy 2015*; p. 149 Precipitation data sourced from Atmosphere, Climate & Environment Information Programme, Department of Environment, Food and Rural Affairs (DEFRA); p. 151 Department of Energy and Climate Change; p.153 ChartsBin statistics collector team 2011, Daily Calorie Intake Per Capita, ChartsBin.com, viewed 24th April, 2017, http://chartsbin.com/view/1150; p.154 WRI Aqueduct, World Resources Institute p.156 AeroFarms, LLC; p.157 NASA International Space Station program and the JSC Earth Science & Remote Sensing Unit, ARES Division, Exploration Integration Science Directorate; p.159 Practical Action Bangladesh; p.160 WRI Aqueduct, World Resources Institute; p. 161 © WaterAid/ Basile Ouedraogo; p 164 Adapted from map by Office of the SNWD Project Commission of the State Council/J.B.; p 166 World Economic Forum and Accenture analysis; p. 167 UN Data; p.169 (top) Source International Energy Agency, 2016; p. 169 (bottom) iStockphoto/dan_prat; p.171 © A Goldstein/ Survival; p. 174 DBEIS Data; p.183 © Bündnis Entwicklung Hilft: WorldRiskReport 2016